When I was young and able
I sat upon the table,
The table broke
And gave me a poke,
When I was young and able.

Cups and saucers
Plates and dishes
My old man wears
Calico breeches.

Eeny, teeny, ether, fether, fip
    Satha, latha, ko, darthur, dick,
Ten-dick, teen-dick, ether-dick, feather-dick,
    Een-bunkin, teen-bunkin, ether-bunkin, fether-bunkin, digit.

My mother said
That the rope must go
Over my head.

Half a pound of twopenny ri
Half a pound of treacle
Penn'orth of spice
To make it nice
Pop goes the weasel.

Dancing Dolly had no sense
For to fiddle for eighteen pence;
All the tunes that she could play,
Were 'Sally get out of the donkey's way'.

Piggy on the railway, picking up the stones
Up came an engine, and broke Piggy's bones.
Oh, said Piggy, that's not fair –
Oh, said the driver, I don't care.

One two three
    Father caught a flea;
He put it in the teapot
    To make a cup of tea.

Up and down the ladder wall,
Penny loaf to feed us all;
A bit for you and a bit for me,
And a bit for all the familee.

The Paint Box

'Cobalt and umber and ultramarine,
Ivory black and emerald green –
What shall I paint to give pleasure to you?'
'Paint for me somebody utterly new.'

'I have painted you tigers in crimson and white.'
'The colours were good and you painted aright.'
'I have painted the cook and a camel in blue
And a panther in purple.' 'You painted them true.

Now mix me a colour that nobody knows,
And paint me a country where nobody goes,
And put in it people a little like you,
Watching a unicorn drinking the dew.'

E. V. Rieu

The *Oxford Treasury of Children's Poems*

*Michael Harrison
and Christopher Stuart-Clark*

Oxford University Press
Oxford   New York   Toronto

Contents

Baby's Drinking Song

Sip a little
Sup a little
From your little
Cup a little
Sup a little
Sip a little
Put it to your
Lip a little
Tip a little
Tap a little
Not into your
Lap or it'll
Drip a little
Drop a little
On the table
Top a little.

James Kirkup

Don't Care

Don't Care didn't care,
    Don't Care was wild:
Don't Care stole plum and pear
    Like any beggar's child.

Don't Care was made to care,
    Don't Care was hung:
Don't Care was put in a pot
    And boiled till he was done.

Anon.

10

Swing, Swing

Swing, swing,
Sing, sing,
Here! my throne and I am a king!
Swing, sing,
Swing, sing,
Farewell, earth, for I'm on the wing!

Low, high,
Here I fly,
Like a bird through sunny sky;
Free, free,
Over the lea,
Over the mountain, over the sea!

Soon, soon,
Afternoon,
Over the sunset, over the moon;
Far, far,
Over all bar,
Sweeping on from star to star!

No, no,
Low, low,
Sweeping daisies with my toe.
Slow, slow,
To and fro,
Slow — slow — slow — slow.

William Allingham

11

Old King Cole

Old King Cole was a merry old soul,
And a merry old soul was he;
He called for his pipe,
And he called for his bowl,
And he called for his fiddlers three.

Then he called for his fifers two,
And they puffed and they blew tootle-oo;
And King Cole laughed as his glass he quaffed,
And his fifers puffed tootle-oo.

Then he called for his drummer boy,
The army's pride and joy,
And the thuds out-rang with a loud bang! bang!
The noise of the noisiest toy.

Then he called for his trumpeters four,
Who stood at his own palace door,
And they played trang-a-tang
Whilst the drummer went bang,
And King Cole he called for more.

He called for a man to conduct,
Who into his bed had been tuck'd,
And he had to get up without bite or sup
And waggle his stick and conduct.

Old King Cole laughed with glee,
Such rare antics to see;
There never was a man in merry England
Who was half as merry as he.

Anon.

The Key of the Kingdom

This is the key of the Kingdom:
In that Kingdom is a city;
In that city is a town;
In that town is a street;
In that street there winds a lane;
In that lane there is a yard;
In that yard there is a house;
In that house there waits a room;
In that room an empty bed,
And on that bed a basket—
A basket of sweet flowers
   Of flowers, of flowers;
    A basket of sweet flowers.

Flowers in a basket;
Basket on the bed;
Bed in the room;
Room in the house;
House in the yard;
Yard in the winding lane;
Lane in the street;
Street in the town;
Town in the city;
City in the Kingdom—
This is the key of the Kingdom.
   Of the Kingdom this is the key.

Anon.

Birthdays

Monday's child is fair of face,
Tuesday's child is full of grace,
Wednesday's child is full of woe,
Thursday's child has far to go,
Friday's child is loving and giving,
Saturday's child works hard for its living;
But the child who is born on the Sabbath day
Is bonny and blithe and good and gay.

Anon.

Sneezing

Sneeze on Monday, sneeze for danger;
Sneeze on Tuesday, miss a stranger;
Sneeze on Wednesday, get a letter;
Sneeze on Thursday, something better;
Sneeze on Friday, sneeze for sorrow
Sneeze on Saturday,
   see your sweetheart tomorrow.

Anon.

Solomon Grundy

Solomon Grundy,
Born on Monday,
Christened on Tuesday,
Married on Wednesday,
Took ill on Thursday,
Worse on Friday,
Died on Saturday,
Buried on Sunday,
So that was the end
   of Solomon Grundy.

Anon.

A Man of Words

A man of words and not of deeds
Is like a garden full of weeds;
And when the weeds begin to grow,
It's like a garden full of snow;
And when the snow begins to fall,
It's like a bird upon the wall;
And when the bird away does fly,
It's like an eagle in the sky;
And when the sky begins to roar,
It's like a lion at the door;
And when the door begins to crack,
It's like a stick across your back;
And when your back begins to smart,
It's like a penknife in your heart;
And when your heart begins to bleed,
You're dead, and dead, and dead indeed.

Anon.

Whether

Whether the weather be fine
Or whether the weather be not
Whether the weather be cold
Or whether the weather be hot –
We'll weather the weather
Whatever the weather
Whether we like it or not!

Anon.

Harum Scarum

I am harum
I disturb the peace
I go around
saying boo to geese

I am scarum
I tell white lies
given half a chance
I would hurt flies

I am harum scarum
a one man gang
diddle dum darum
bang bang bang

Roger McGough

16

Simple Simon

Simple Simon met a pieman
Going to the fair;
Says Simple Simon to the pieman,
Let me taste your ware.

Says the pieman to Simple Simon,
Show me first your penny;
Says Simple Simon to the pieman,
Indeed I have not any.

Simple Simon went a-fishing,
For to catch a whale;
All the water he had got
Was in his mother's pail.

Simple Simon went a-hunting,
For to catch a hare;
He rode a goat about the streets,
But couldn't find one there.

He went to catch a dickey bird,
And thought he could not fail,
Because he'd got a little salt
To put upon its tail.

He went to shoot a wild duck,
But wild duck flew away;
Says Simon, I can't hit him,
Because he will not stay.

He went to ride a spotted cow,
That had a little calf;
She threw him down upon the ground,
Which made the people laugh.

Once Simon made a great snowball,
And brought it in to roast;
He laid it down before the fire,
And soon the ball was lost.

He went to try if cherries ripe
Did grow upon a thistle;
He pricked his finger very much
Which made poor Simon whistle.

He went for water in a sieve,
But soon it all ran through;
And now poor Simple Simon
Bids you all adieu.

Anon.

Two Witches

There was a witch
The witch had an itch
The itch was so itchy it
Gave her a twitch.

Another witch
Admired the twitch
So she started twitching
Though she had no itch.

Now both of them twitch
So it's hard to tell which
Witch has the itch and
Which witch has the twitch.

Charles Reznikoff

My Mother Said . . .

My Mother said,
I never should,
Play with the gipsies
In the wood;
If I did, she would say,
'Naughty little girl to disobey.
Your hair won't curl,
And your eyes won't shine,
You gipsy girl,
You shan't be mine!'

And my father said that if I did
He'd rap my head with the teapot lid.
The wood was dark, the grass was green,
In came Sally with a tambourine.
I went to sea – no ship to get across;
I paid ten shillings for a blind white horse;
I up on his back and was off in a crack,
Sally, tell my mother I shall never come back!

Anon.

There was an Orchestra

There was an orchestra — Bingo-Bango
Playing for us to dance the tango
And the people all clapped as we arose
For her sweet face and my new clothes.

F. Scott Fitzgerald

Jeremiah Obadiah

Jeremiah Obadiah, puff, puff, puff,
When he gives his messages he snuffs, snuffs, snuffs,
When he goes to school by day, he roars, roars, roars,
When he goes to bed at night he snores, snores, snores,
When he goes to Christmas treat he eats plum-duff,
Jeremiah Obadiah, puff, puff, puff.

Anon.

Picnic

Ella, fell a
Maple tree.
Hilda, build a
Fire for me.

Teresa, squeeze a
Lemon, so
Amanda, hand a
Plate to Flo.

Nora, pour a
Cup of tea.
Fancy, Nancy,
What a spree!

Hugh Lofting

Mr Bidery's Spidery Garden

Poor old Mr Bidery.
His garden's awfully spidery:
Bugs use it as a hidery.

In April it was seedery,
By May a mass of weedery;
And oh, the bugs! How greedery.

White flowers out or buddery,
Potatoes made it spuddery,
And when it rained, what muddery!

June days grow long and shaddery;
Bullfrog forgets his taddery;
The spider legs his laddery.

With cabbages so odoury,
Snapdragon soon explodery,
At twilight all is toadery.

Young corn still far from foddery
No sign of goldenrodery,
Yet feeling low and doddery.

Is poor old Mr Bidery,
His garden lush and spidery,
His apples green, not cidery.

Pea-picking *is* so poddery!

David McCord

I Saw

I saw a peacock with a fiery tail
I saw a blazing comet drop down hail
I saw a cloud with ivy circled round
I saw a sturdy oak creep on the ground
I saw an ant swallow up a whale
I saw a raging sea brim full of ale
I saw a Venice glass sixteen foot deep
I saw a well full of men's tears that weep
I saw their eyes all in a flame of fire
I saw a house as big as the moon and higher
I saw the sun even in the midst of night
I saw the man that saw this wondrous sight.

I saw a fishpond all on fire
I saw a house bow to a squire
I saw a parson twelve feet high
I saw a cottage near the sky
I saw a balloon made of lead
I saw a coffin drop down dead
I saw a sparrow run a race
I saw two horses making lace
I saw a girl just like a cat
I saw a kitten wear a hat
I saw a man who saw these too,
And says, though strange, they all are true.

Anon.

Peter Piper

Peter Piper picked a peck of pickled pepper;
Did Peter Piper pick a peck of pickled pepper?
If Peter Piper picked a peck of pickled pepper,
Where's the peck of pickled pepper Peter Piper picked?

Anon.

She Sells Sea-shells

She sells sea-shells on the sea shore;
The shells that she sells are sea-shells I'm sure.
So if she sells sea-shells on the sea shore,
I'm sure that the shells are sea-shore shells.

Anon.

Betty Botter

Betty Botter bought some butter,
But, she said, this butter's bitter;
If I put it in my batter,
It will make my batter bitter,
But a bit of better butter
Will make my batter better.
So she bought a bit of butter
Better than her bitter butter,
And she put it in her batter,
And it made her batter better,
So 'twas a better Betty Botter
Bought a bit of better butter.

Anon.

Tailor

I saw a little Tailor sitting stitch, stitch, stitching
Cross-legged on the floor of his kitch, kitch, kitchen.
His thumbs and his fingers were so nim, nim, nimble
With his wax and his scissors and his thim, thim, thimble.

His silk and his cotton he was thread, thread, threading
For a gown and a coat for a wed, wed, wedding,
His needle flew as swift as a swal, swal, swallow,
And his spools and his reels had to fol, fol, follow.

He hummed as he worked a merry dit, dit, ditty:
'The Bride is as plump as she's pret, pret, pretty,
I wouldn't have her taller or short, short, shorter,
She can laugh like the falling of wat, wat, water.

'She can put a cherry-pie, togeth, geth, gether,
She can dance as light as a feath, feath, feather,
She can sing as sweet as a fid, fid, fiddle,
And she's only twenty inches round the mid, mid, middle.'

The happy little Tailor went on stitch, stitch, stitching
The black and the white in his kitch, kitch, kitchen.
He will wear the black one, she will wear the white one,
And the knot the Parson ties will be a tight, tight, tight one.

Eleanor Farjeon

Potato Clock

A potato clock, a potato clock
   Has anybody got a potato clock?
A potato clock, a potato clock
   Oh where can I find a potato clock?

I went down to London the other day
Found myself a job with a lot of pay
Carrying bricks on a building site
From early in the morning till late at night.

No one here works as hard as me
I never even break for a cup of tea
My only weakness, my only crime
Is that I can never get to work on time.

A potato clock, a potato clock
   Has anybody got a potato clock?
A potato clock, a potato clock
   Oh where can I find a potato clock?

I arrived this morning half an hour late
The foreman came up in a terrible state
'You've got a good job, but you'll lose it, cock,
If you don't get up at eight o'clock.'

Up at eight o'clock, up at eight o'clock
   Has anybody got up at eight o'clock?
Up at eight o'clock, up at eight o'clock
   Oh where can I find up at eight o'clock?

Roger McGough

A Good Play

We built a ship upon the stairs
All made of the back-bedroom chairs,
And filled it full of sofa pillows
To go a-sailing on the billows.

We took a saw and several nails,
And water in the nursery pails;
And Tom said, 'Let us also take
An apple and a slice of cake';
Which was enough for Tom and me
To go a-sailing on, till tea.

We sailed along for days and days,
And had the very best of plays;
But Tom fell out and hurt his knee,
So there was no one left but me.

Robert Louis Stevenson

Through Nurseryland

Now, rocking horse! rocking horse! where shall we go?
The world's such a very big place, you must know,
That to see all its wonders, the wiseacres say,
'Twould take us together a year and a day.

Suppose we first gallop to Banbury Cross,
To visit that lady upon a white horse,
And see if it's true that her fingers and toes
Make beautiful music, wherever she goes.

Then knock at the door of the Old Woman's Shoe,
And ask if her wonderful house is on view,
And peep at the children, all tucked up in bed,
And beg for a taste of the broth without bread.

On poor Humpty-Dumpty we'll certainly call,
Perhaps we might help him to get back on his wall;
Spare two or three minutes to comfort the Kits
Who've been kept without pie, just for losing their mits.

A rush to Jack Horner's, then down a steep hill,
Not over and over, like poor Jack and Jill!
So, rocking horse! rocking horse! scamper away,
Or we'll never get back in a year and a day.

Anon.

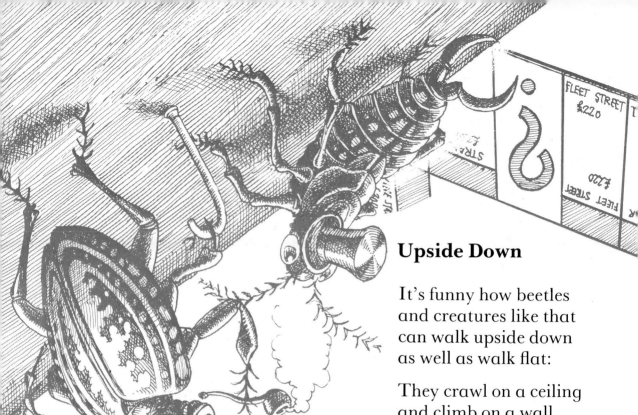

Upside Down

It's funny how beetles
and creatures like that
can walk upside down
as well as walk flat:

They crawl on a ceiling
and climb on a wall
without any practice
or trouble at all,

While I have been trying
for a year (maybe more)
and still I can't stand
with my head on the floor.

Aileen Fisher

Truth

Sticks and stones may break my bones,
but words can also hurt me.
Stones and sticks break only skin,
while words are ghosts that haunt me.

Slant and curved the word-swords fall
to pierce and stick inside me.
Bats and bricks may ache through bones,
but words can mortify me.

Pain from words has left its scar
on mind and heart that's tender.
Cuts and bruises now have healed;
it's words that I remember.

Barrie Wade

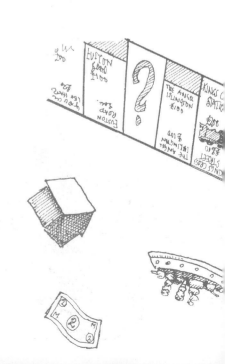

It's a Bit Rich

Playing Monopoly's
Really my scene.
I hang on to houses
And play very mean.

I take all the money.
There's often a stack.
I'm not very pleasant
When giving it back.

I'm harsh as a landlord.
I've nothing for sale.
I'm buying your station.
You're going to jail.

My fistful of money —
It seems such a shame
When bedtime arrives
And it's only a game.

Max Fatchen

29

The Black Pebble

There went three children down to the shore
   Down to the shore and back;
There was skipping Susan and bright-eyed Sam
   And little scowling Jack.

Susan found a white cockle-shell,
   The prettiest ever seen,
And Sam picked up a piece of glass
   Rounded and smooth and green.

But Jack found only a plain black pebble
   That lay by the rolling sea,
And that was all that ever he found;
   So back they went all three.

The cockle-shell they put on the table,
   The green glass on the shelf,
But the little black pebble that Jack had found
   He kept it for himself.

James Reeves

The Tree in the Garden

There's a tree out in our garden which is very nice to climb,
And I often go and climb it when it's fine in summer time,
And when I've climbed right up it I pretend it's not a tree
But a ship in which I'm sailing, far away across the sea.

Its branches are the rigging and the grass so far below
I make believe's the ocean over which my ship must go;
And when the wind is blowing then I really seem to be
A-sailing, sailing, sailing, far away across the sea.

Then I hunt for desert islands and I very often find
A chest stuffed full of treasure which some pirate's left behind –
My good ship's hold is filled with gold – it all belongs to me –
For I've found it when I'm sailing far away across the sea.

It's a lovely game to play at – though the tree trunk's rather green,
Still, when I'm in my bath at night I always come quite clean.
And so through all the summer, in my good ship Treasure-Tree,
I shall often go a-sailing far away across the sea.

Christine Chaundler

The Friendly Cinnamon Bun

Shining in his stickiness and glistening with honey,
Safe among his sisters and his brothers on a tray,
With raisin eyes that looked at me as I put down my money,
There smiled a friendly cinnamon bun, and this I heard him say:

'It's a lovely, lovely morning, and the world's a lovely place;
I know it's going to be a lovely day.
I know we're going to be good friends; I like your honest face;
Together we might go a long, long way.'

The baker's girl rang up the sale, 'I'll wrap your bun,' said she.
'Oh no, you needn't bother,' I replied.
I smiled back at that cinnamon bun and ate him, one two three,
And walked out with his friendliness inside.

Russell Hoban

Waiting

Waiting, waiting, waiting
   For the party to begin;
Waiting, waiting, waiting
   For the laughter and din;
Waiting, waiting, waiting
   With hair just so
And clothes trim and tidy
   From top-knot to toe.
The floor is all shiny,
   The lights are ablaze;
There are sweetmeats in plenty
   And cakes beyond praise.
Oh the games and dancing,
   The tricks and the toys,
The music and the madness
   The colour and noise!
Waiting, waiting, waiting
   For the first knock on the door –
Was ever such waiting,
   Such waiting before?

James Reeves

Copycat

Copycat, copycat,
Shadow's a copycat!

Out in the sun
Whenever I run,
It runs.
Whenever I twirl,
It twirls.
I curl up small.
It curls up small.
I stand up tall.
It stands up tall.

Copycat, copycat,
Shadow's a copycat.

Whenever I hide,
It hides.
I spread out wide.
It spreads out wide.
I pat my head.
It pats its head.
I fall down dead.
It falls down dead.

But when I go inside
   to stay,
Copycat, copycat
   goes away!

Robert Heidbreder

My Playmate

I often wonder how it is
   That on a rainy day,
A little boy, just like myself,
   Comes out with me to play.

And we step in all the puddles
   When walking into town,
But though I stand the right way up,
   He's always upside-down.

I have to tread upon his feet,
   Which is a sorry sight,
With my right foot on his left foot,
   My left foot on his right.

I really wish he'd talk to me,
   He seems so very kind,
For when I look and smile at him
   He does the same, I find.

But I never hear him speaking,
   So surely he must be
In some strange land the other side,
   Just opposite to me.

Mary I. Osborn

My Shadow

I have a little shadow that goes in and out with me,
And what can be the use of him is more than I can see.
He is very, very like me from the heels up to the head;
And I see him jump before me, when I jump into my bed.

The funniest thing about him is the way he likes to grow —
Not at all like proper children, which is always very slow;
For he sometimes shoots up taller like an india-rubber ball,
And he sometimes gets so little that there's none of him at all.

He hasn't got a notion of how children ought to play,
And can only make a fool of me in every sort of way.
He stays so close beside me, he's a coward you can see;
I'd think shame to stick to nursie as that shadow sticks to me!

One morning, very early, before the sun was up,
I rose and found the shining dew on every buttercup;
But my lazy little shadow, like an arrant sleepy-head,
Had stayed at home behind me and was fast asleep in bed.

Robert Louis Stevenson

Egg-Thoughts

Soft-boiled
I do not like the way you slide,
I do not like your soft inside,
I do not like you many ways,
And I could do for many days
Without a soft-boiled egg.

Sunny-Side-Up
With their yolks and whites all runny
They are looking at me funny.

Sunny-Side-Down
Lying face-down on the plate
On their stomachs there they wait.

Poached
Poached eggs on toast, why do you shiver
With such a funny little quiver?

Scrambled
I eat as well as I am able,
But some falls underneath the table.

Hard-boiled
With so much suffering today
Why do them any other way?

Russell Hoban

The Toaster

A silver-scaled Dragon with jaws flaming red
Sits at my elbow and toasts my bread.
I hand him fat slices, and then, one by one,
He hands them back when he sees they are done.

William Jay Smith

Boiling an Egg

The egg I have chosen is sandy brown
And lies in the pan like a stone in a stream.
The water begins to bubble and steam
And the egg bumps up and down.

From the bottom the bubbles rise and swell
Like treasure-seeking divers gasping for breath.
When the egg is done I shall break its shell
And open it like a treasure chest.

Stanley Cook

Seven Activities for a Young Child

Turn on the tap for straight and silver water in the sink,
Cross your finger through
The sleek thread falling   *One*

Spread white sandgrains on a tray,
And make clean furrows with a bent stick
To stare for a meaning   *Two*

Draw some clumsy birds on yellow paper,
Confronting each other and as if to fly
Over your scribbled hill   *Three*

Cut rapid holes into folded paper, look
At the unfolded pattern, look
Through the unfolded pattern   *Four*

Walk on any square stone of the pavement,
Or on any crack between, as long
As it's with no one or with someone   *Five*

Throw up a ball to touch the truest brick
Of the red-brick wall,
Catch it with neat, cupped hand   *Six*

Make up in your head a path, and name it,
Name where it will lead you,
Walk towards where it will lead you   *Seven*

One, two, three, four, five, six, seven:
Take-up-the-rag-doll-quietly-and-sing-her-to-sleep.

Alan Brownjohn

New Shoes

Buying new shoes
takes so long.
When the colour is right
the size is wrong.

The lady asks
How does it fit?
I say to Mum
Pinches a bit.

But that's not true
It's just because
I don't want the brown
I prefer the blue.

The lady goes inside
brings another size
this time the blue.
Not too big. Not too tight.

As you guessed
Just right, just right.
Mum says, 'The blue will do.'
And I agree. Don't you?

John Agard

Supermarket

I'm
lost
among a
maze of cans
behind a pyramid
of jams, quite near
asparagus and rice,
close to the Oriental spice,
and just before sardines.

I hear my mother calling, 'Joe.
Where are you, Joe?
Where did you
Go?' And I reply in a voice conceale[d]
among the candied orange peel,
and packs of Chocolate Dreams.

'I
hear
you, Mother
dear, I'm here –

quite near the ginger ale
and beer, and lost among a
maze
of cans
behind a
pyramid of jams
quite near asparagus
and rice, close to the
Oriental spice, and just
before sardines.'

But
still
my mother
calls me, 'Joe!
Where are you, Joe?
Where did you go?

'Somewhere
around asparagus
that's in a sort of
broken glass,
        beside a kind of m-
                ess-
                    y jell
        that's near a tower of cans that
                                            f
                                            e
                                            l
                                            l

        and squashed the Chocolate Dreams.

Felice Holman

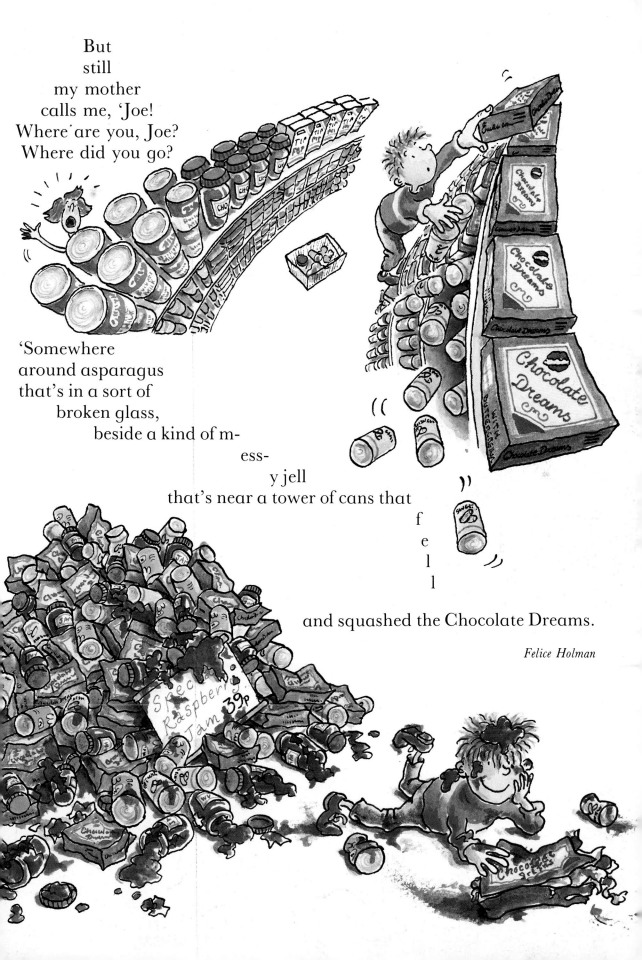

Meet-on-the-Road

'Now pray, where are you going, child?'
   said Meet-on-the-Road.
'To school, sir, to school, sir,'
   said Child-as-it-Stood.
'What have you got in your basket, child?'
   said Meet-on-the-Road.
'My dinner, sir, my dinner, sir,'
   said Child-as-it-Stood.
'What have you for your dinner, child?'
   said Meet-on-the-Road.

'Some pudding, sir, some pudding, sir,'
   said Child-as-it-Stood.
'Oh then, I pray, give me a share,'
   said Meet-on-the-Road.
'I've little enough for myself, sir,'
   said Child-as-it-Stood.
'What have you got that cloak on for?'
   said Meet-on-the-Road.
'To keep the wind and the cold from me,'
   said Child-as-it-Stood.

'I wish the wind would blow through you,'
   said Meet-on-the-Road.
'Oh, what a wish! Oh, what a wish!'
   said Child-as-it-Stood.
'Pray, what are those bells ringing for?'
   said Meet-on-the-Road.
'To ring bad spirits home again,'
   said Child-as-it-Stood.
'Oh, then I must be going, child!'
   said Meet-on-the-Road.
'So fare you well, so fare you well,'
   said Child-as-it-Stood.

Anon.

A Boy's Song

Where the pools are bright and deep,
Where the grey trout lies asleep,
Up the river and o'er the lea,
That's the way for Billy and me.

Where the blackbird sings the latest,
Where the hawthorn blooms the sweetest,
Where the nestlings chirp and flee,
That's the way for Billy and me.

Where the mowers mow the cleanest,
Where the hay lies thick and greenest;
There to trace the homeward bee,
That's the way for Billy and me.

Where the hazel bank is steepest,
Where the shadow falls the deepest,
Where the clustering nuts fall free,
That's the way for Billy and me.

Why the boys should drive away
Little sweet maidens from their play,
Or love to banter and fight so well,
That's the thing I never could tell.

But this I know, I love to play
Through the meadow, among the hay;
Up the water and o'er the lea,
That's the way for Billy and me.

James Hogg

The Sugar-Plum Tree

Have you ever heard of the Sugar-PlumTree?
'Tis a marvel of great renown!
It blooms on the shore of the Lollipop sea
In the garden of Shut-Eye Town;
The fruit that it bears is so wondrously sweet
(As those who have tasted it say)
That good little children have only to eat
Of that fruit to be happy next day.

When you've got to the tree, you would have a hard time
To capture the fruit which I sing;
The tree is so tall that no person could climb
To the boughs where the sugar-plums swing!
But up in that tree sits a chocolate cat,
And a gingerbread dog prowls below –
And this is the way you contrive to get at
Those sugar-plums tempting you so:

You say but the word to that gingerbread dog
And he barks with such terrible zest
That the chocolate cat is at once all agog,
As her swelling proportions attest.

And the chocolate cat goes cavorting around
From this leafy limb unto that,
And the sugar-plums, tumble, of course, to the ground –
Hurrah for that chocolate cat!

There are marshmallows, gumdrops, and peppermint canes,
With stripings of scarlet or gold,
And you carry away of the treasure that rains
As much as your apron can hold!
So come, little child, cuddle closer to me
In your dainty white nightcap and gown,
And I'll rock you away to that Sugar-Plum Tree
In the garden of Shut-Eye Town.

Eugene Field

One, Two, Three —

If you don't put your shoes on before I count fifteen then
we won't go to the woods to climb the chestnut, one
    But I can't find them.
Two.
    I can't
They're under the sofa, three
    No. O yes
Four five six
    Stop – they've got knots they've got knots
You should untie the laces when you take your shoes off, seven
    Will you do one shoe while I do the other then?
Eight, but that would be cheating
    Please
All right
    It always . . .
Nine
    It always sticks – I'll use my teeth
Ten
    It won't it won't. It has – look
Eleven
    I'm not wearing any socks.
Twelve
    Stop counting stop counting. Mum, where are my socks, mum?
They're in your shoes. Where you left them.
    I didn't.
Thirteen
    O, they're inside out and upside down and bundled up
Fourteen
    Have you done the knot on the shoe you were . . .
Yes, put it on the right foot
    But socks don't have a right and wrong foot
The shoes silly. Fourteen and a half.

I am I am. Wait
Don't go to the woods without me
Look that's one shoe already
Fourteen and threequarters
   There
You haven't tied the bows yet
   We could do them on the way there
No we won't. Fourteen and seven eighths
   Help me then.
   You know I'm not fast at bows
Fourteen and fifteen sixteeeenths
   A single bow is alright isn't it?
Fifteen. We're off.
   See I did it.
   Didn't I?

Michael Rosen

After a Bath

After my bath
I try, try, try
to wipe myself
till I'm dry, dry, dry.

Hands to wipe
and fingers and toes
and two wet legs
and a shiny nose.

Just think how much
less time I'd take
if I were a dog
and could shake,
   shake, shake.

Aileen Fisher

Wonders of Nature

My Grandmother said, 'Now isn't it queer,
That boys must whistle and girls must sing?
But that's how 'tis!' I heard her say –
'The same tomorrow as yesterday.'

Grandmother said, when I asked her why
Girls couldn't whistle the same as I,
'Son you know it's a natural thing –
Boys just whistle, and girls just sing.'

Anon.

Squeezes

We love to squeeze bananas,
We love to squeeze ripe plums,
And when they are feeling sad
We love to squeeze our mums.

Brian Patten

Ask Mummy Ask Daddy

When I ask Daddy
Daddy says ask Mummy.

When I ask Mummy
Mummy says ask Daddy.
I don't know where to go.

Better ask my teddy
He never says no.

John Agard

Advice to a Child

Set your fir-tree
In a pot;
Needles green
Is all it's got.
Shut the door
And go away,
And so to sleep
Till Christmas Day.
In the morning
Seek your tree,
And you shall see
What you shall see.

Hang your stocking
By the fire,
Empty of
Your heart's desire;

Up the chimney
Say your say,
And so to sleep
Till Christmas Day.
In the morning
Draw the blind,
And you shall find
What you shall find.

Eleanor Farjeon

Nobody

Nobody loves me,
Nobody cares,
Nobody picks me peaches and pears.
Nobody offers me candy and Cokes,
Nobody listens and laughs at my jokes.
Nobody helps when I get in a fight,
Nobody does all my homework at night.
Nobody misses me,
Nobody cries,
Nobody thinks I'm a wonderful guy.
So if you ask me who's my best friend, in a whiz,
I'll stand up and tell you that *Nobody* is.
But yesterday night I got quite a scare,
I woke up and Nobody just *wasn't there*.
I called out and reached out for Nobody's hand,
In the darkness where Nobody usually stands.
Then I poked through the house, in each cranny and nook,
But I found *somebody* each place that I looked.
I searched till I'm tired, and now with the dawn,
There's no doubt about it —
Nobody's *gone*!

Shel Silverstein

I'm Nobody! Who are You?

I'm Nobody! Who are you?
Are you – Nobody – Too?
Then there's a pair of us!
Don't tell! they'd banish us you know!

How dreary – to be – Somebody!
How public – like a Frog –
To tell your name – the livelong June –
To an admiring Bog!

Emily Dickinson

How I See it

Some say the world's
A hopeless case:
A speck of dust
In all that space.
It's certainly
A scruffy place.
Just one hope
For the human race
That I can see;
Me. I'm
ACE!

Kit Wright

Me – Pirate

If ever I go to sea,
I think I'll be a pirate:
I'll have a treasure-ship in tow
And a man-of-war to fire at.

With a cutlass at my belt,
And a pistol in my hand,
I'll nail my Crossbones to the mast
And sail for a foreign land.

And when we reach that shore,
We'll beat our battle-drum
And fire a salute of fifteen guns
To tell them we have come.

We'll fight them all day long;
We'll seize their chests of gold,
Their diamonds, coins and necklaces,
And stuff them in our hold.

A year and a day at home,
Then off on the waves again –
Lord of the Caribbean Seas
King of the Spanish Main!

Clive Sansom

If Pigs could Fly

If Pigs could fly, I'd fly a pig
To foreign countries small and big –
   To Italy and Spain,
To Austria, where cowbells ring,
To Germany, where people sing –
   And then come home again.

I'd see the Ganges and the Nile:
I'd visit Madagascar's isle,
   And Persia and Peru.
People would say they'd never seen
So odd, so strange an air-machine
   As that on which I flew.

Why, everyone would raise a shout
To see his trotter and his snout
   Come floating from the sky;
And I would be a famous star
Well known in countries near and far –
   If only pigs could fly!

James Reeves

I Don't Believe in Human-Tales

I don't believe there's such a thing
As nasty little boys.
I think that someone dreams them up;
It's one of the gnome-up's ploys.

I don't believe in super-stores
Where bits of wings and legs
Of harmless little chickens
Are sold in plastic bags.

I'm sure it is just rubbish
That we get turned to stone
If we go near garden ponds
When we're playing out alone.

I'm sure gnome-ups invent these things
To scare us little gnomes
So we'll never leave the forests
Under which we will have our homes.

Brian Patten

The Teasing Toads

'I hate your laces
I undo them.'
'I love your buttons
I eat them.'

'I am soap-sting
I get in your eye
and make you cry.'

'I'm the one
who burns your tongue.'
'I get up your nose
and make it run.'

'I am boot-bug, living in your boots
when you tug – I tug
I don't let go.'

'And when at night
into bed you creep
I'll be there to ruck your sheets.'

'Soap-stings, boot-bugs
long-distance nose runners are we.
Every way any time,
we tease, we tease.'

Michael Rosen

55

Amanda!

Don't bite your nails, Amanda!
Don't hunch your shoulders, Amanda!
Stop that slouching and sit up straight,
Amanda!

(There is a languid, emerald sea,
where the sole inhabitant is me –
a mermaid, drifting blissfully.)

Did you finish your homework, Amanda?
Did you tidy your room, Amanda?
I thought I told you to clean your shoes,
Amanda!

(I am an orphan, roaming the street.
I pattern soft dust with my hushed, bare feet.
The silence is golden, the freedom is sweet.)

Don't eat that chocolate, Amanda!
Remember your acne, Amanda!
Will you please look at me when I'm speaking to you,
Amanda!

(I am Rapunzel, I have not a care;
life in a tower is tranquil and rare;
I'll certainly *never* let down my bright hair!)

Stop that sulking at once, Amanda!
You're always so moody, Amanda!
Anyone would think that I nagged at you,
Amanda!

Robin Klein

I Wonder

I wonder why the grass is green,
And why the wind is never seen?

Who taught the birds to build a nest,
And told the trees to take a rest?

O, when the moon is not quite round,
Where can the missing bit be found?

Who lights the stars, when they blow out,
And makes the lightning flash about?

Who paints the rainbow in the sky,
And hangs the fluffy clouds so high?

Why is it now, do you suppose,
That Dad won't tell me, if he knows?

Jeannie Kirby

Chivvy

Grown-ups say things like:
Speak up
Don't talk with your mouth full
Don't stare
Don't point
Don't pick your nose

Sit up
Say please
Less noise
Shut the door behind you
Don't drag your feet
Haven't you got a hankie?
Take your hands out of
    your pockets

Pull your socks up
Stand up straight
Say thank you
Don't interrupt
No one thinks you're funny
Take your elbows off the table

Can't you make your *own*
mind up about anything?

Michael Rosen

The Frogologist

I hate it when grown-ups say,
'What do you want to be?'
I hate the way they stand up there
And talk down to me.

I say:
'I want to be a frogologist
And study the lives of frogs,
I want to know their habitat
And crawl about in bogs,
I want to learn to croak and jump
And catch flies with my tongue
And will they please excuse me 'cause
Frogologists start quite young.'

Brian Patten

Grown-ups

Talking for yonks about babies and
        washing machines
telling us off for scratching
throwing our best toys
in the bin without permission

picking fluff from our jerseys
switching cartoons off for The News
talking about us – 'he' does this
'he' does that – while we're still there
asking us all those times what we want for supper
when they should KNOW

why can't they DO something

figure of eights on one wheel
abseiling out of the bedroom window
holing up in the chimney
trip-wiring the hall
damming the gutters
holding your breath for how many minutes
painting the wall with your toes
digging the lawn for moles?

Geoffrey Holloway

Dad and the Cat and the Tree

This morning a cat got
Stuck in our tree.
Dad said, 'Right, just
Leave it to me.'

The tree was wobbly,
The tree was tall.
Mum said, 'For goodness'
Sake don't fall!'

'Fall?' scoffed Dad,
'A climber like me?
Child's play, this is!
You wait and see.'

He got out the ladder
From the garden shed.
It slipped. He landed
In the flower bed.

'Never mind', said Dad,
Brushing the dirt
Off his hair and his face
And his trousers and his shirt,

'We'll try Plan B. Stand
Out of the way!'
Mum said, 'Don't fall
Again, O.K.?'

'Fall again?' said Dad.
'Funny joke!'
Then he swung himself up
On a branch. It broke.

Dad landed *wallop*
Back on the deck.
Mum said, 'Stop it,
You'll break your neck!'

'Rubbish!' said Dad.
'Now we'll try Plan C.
Easy as winking
To a climber like me!'

Then he climbed up high
On the garden wall.
Guess what?
He *didn't fall*!

He gave a great leap
And he landed flat
In the crook of the tree-trunk –
Right on the cat!

The cat gave a yell
And sprang to the ground,
Pleased as Punch to be
Safe and sound.

So it's smiling and smirking,
Smug as can be,
But poor old Dad's
Still

Stuck
Up
The
Tree!

Kit Wright

Disobedience

James James
Morrison Morrison
Weatherby George Dupree
Took great
Care of his Mother,
Though he was only three.
James James
Said to his Mother,
'Mother,' he said, said he;
'You must never go down to the end of the town,
  If you don't go down with me.'

James James
Morrison's Mother
Put on a golden gown,
James James
Morrison's Mother
Drove to the end of the town.
James James
Morrison's Mother
Said to herself, said she:
'I can get right down to the end of the town and be
  back in time for tea.'

King John
Put up a notice,
'LOST or STOLEN or STRAYED!
JAMES JAMES
MORRISON'S MOTHER
SEEMS TO HAVE BEEN MISLAID
LAST SEEN
WANDERING VAGUELY:
QUITE OF HER OWN ACCORD,
SHE TRIED TO GET DOWN TO THE END OF
  THE TOWN – **FORTY SHILLINGS REWARD!**

James James
Morrison Morrison
(Commonly known as Jim)
Told his
Other relations
Not to go blaming *him*.
James James
Said to his Mother,
'Mother,' he said, said he:
'You must *never* go down to the end of the town
    without consulting me.'

James James
Morrison's mother
Hasn't been heard of since.
King John
Said he was sorry,
So did the Queen and Prince.
King John
(Somebody told me)
Said to a man he knew:
'If people go down to the end of the town, well, what
    can *anyone do*?'

(*Now then, very softly*)
            J. J.
            M. M.
            W. G. Du P.
            Took great
            C/o his M*****
            Though he was only 3.
            J. J.
            Said to his M*****
            'M*****,' he said, said he:
'You-must-never-go-down-to-the-end-of-the-town-if-
            you-don't go-down-with ME!'

A. A. Milne

Bring on the Clowns

Bring on the clowns!
Bring on the clowns!
Clowns wearing knickers
and clowns
wearing gowns.

Tall clowns and short clowns and skinny and fat,
a flat-footed clown with a jumping-jack hat.
A clown walking under a portable shower,
getting all wet just to water a flower.
A barefoot buffoon with balloons on his toes,
a clown with a polka-dot musical nose.
Clowns wearing teapots and clowns sporting plumes
a clown with a tail made of brushes and brooms.

A balancing clown on a wobbly wheel,
Seventeen clowns in an automobile.
Two jesters on pogo sticks dressed up in kilts,
pursuing a prankster escaping on stilts.
A sad looking clown with a face like a tramp,
a clown with his stomach lit up like a lamp.
How quickly a clown can coax smiles out of frowns
Make way for the merriment . . .
bring on the clowns.

Jack Prelutsky

Sing a Song of People

Sing a song of people
    Walking fast or slow;
People in the city
    Up and down they go.

People on the sidewalk,
People on the bus;
People passing, passing,
In back and front of us.
People on the subway
Underneath the ground;
People riding taxis
Round and round and round.

People with their hats on,
Going in the doors;
People with umbrellas
When it rains and pours.
People in tall buildings
And in stores below;
Riding elevators
Up and down they go.

People walking singly,
People in a crowd;
People saying nothing,
People talking loud.
People laughing, smiling,
Grumpy people too;
People who just hurry
And never look at you!

Sing a song of people
    Who like to come and go;
Sing of city people
    You see but never know!

Lois Lenski

Poor Mrs Prior

Poor Mrs Prior.
Oh it was so tragical;
She went down in the magical
Washing machine.
She thought she'd come out cleaner,
And youthfuller and leaner,
Now nobody has seen her
Since Monday afternoon.
    Poor Mrs Prior,
    Oh how it must try her.
    She won't be getting drier,
    Wherever she may be.

Gerda Mayer

Lollipop Lady

Lollipop lady,
lollipop lady,
wave your magic stick
and make the traffic
stop a while
so we can cross the street.

Trucks and cars
rushing past
have no time for little feet.
They hate to wait
especially when late
but we'll be late too
except for you.

So lollipop lady,
lollipop lady,
in the middle of the street
wave your magic stick
and make the traffic
give way to little feet.

John Agard

Old Mrs Thing-um-e-bob

Old Mrs Thing-um-e-bob,
   Lives at you-know-where,
Dropped her what-you-may-call-it down
   The well of the kitchen stair.

'Gracious me!' said Thing-um-e-bob,
   'This don't look too bright.
I'll ask old Mr What's-his-name
   To try and put it right.'

Along came Mr What's-his-name,
   He said, 'You've broke the lot!
I'll have to see what I can do
   With some of the you-know-what.'

So he gave the what-you-may-call-it a pit
   And he gave it a bit of a pat,
And he put it all together again
   With a little of this and that.

And he gave the what-you-may-call-it a dib
   And he gave it a dab as well
When all of a sudden he heard a note
   As clear as any bell.

'It's as good as new!' cried What's-his-name.
   'But please remember, now,
In future Mrs Thing-um-e-bob
   'You'll have to go you-know-how.'

Charles Causley

Colonel Fazackerley

Colonel Fazackerley Butterworth-Toast
Bought an old castle complete with a ghost,
But someone or other forgot to declare
To Colonel Fazack that the spectre was there.

On the very first evening, while waiting to dine,
The Colonel was taking a fine sherry wine,
When the ghost, with a furious flash and a flare,
Shot out of the chimney and shivered, 'Beware!'

Colonel Fazackerley put down his glass
And said, 'My dear fellow, that's really first class!
I just can't conceive how you do it at all.
I imagine you're going to a Fancy Dress Ball?'

At this, the dread ghost gave a withering cry.
Said the Colonel (his monocle firm in his eye),
'Now just how you do it I wish I could think.
Do sit down and tell me, and please have a drink.'

The ghost in his phosphorous cloak gave a roar
And floated about between ceiling and floor.
He walked through a wall and returned through a pane
And backed up the chimney and came down again.

Said the Colonel, 'With laughter I'm feeling quite weak!
(As trickles of merriment ran down his cheek).
'My house-warming party I hope you won't spurn.
You *must* say you'll come and you'll give us a turn!'

At this, the poor spectre – quite out of his wits –
Proceeded to shake himself almost to bits.
He rattled his chains and he clattered his bones
And he filled the whole castle with mumbles and moans.

But Colonel Fazackerley, just as before,
Was simply delighted and called out, 'Encore!'
At which the ghost vanished, his efforts in vain,
And never was seen at the castle again.

'Oh dear, what a pity!' said Colonel Fazack.
'I don't know his name, so I can't call him back.'
And then with a smile that was hard to define,
Colonel Fazackerley went in to dine.

Charles Causley

The King's Breakfast

The King asked
The Queen, and
The Queen asked
The Dairymaid:
'Could we have some butter for
The Royal slice of bread?'
The Queen asked
The Dairymaid,
The Dairymaid
Said, 'Certainly,
I'll go and tell
The cow
Now
Before she goes to bed.'

The Dairymaid
She curtsied,
And went and told
The Alderney:
'Don't forget the butter for
The Royal slice of bread.'
The Alderney
Said sleepily:
'You'd better tell
His Majesty
That many people nowadays
Like marmalade
Instead.'

The Dairymaid
Said, 'Fancy!'
And went to
Her Majesty.
She curtsied to the Queen, and
She turned a little red:

'Excuse me,
Your Majesty
For taking of
The liberty,
But marmalade is tasty, if
It's very
Thickly
Spread.'

The Queen said
'Oh!'
And went to
His Majesty:
'Talking of the butter for
The royal slice of bread,
Many people
Think that
Marmalade
Is nicer.
Would you like to try a little
Marmalade
Instead?'

The King said,
'Bother!'
And then he said,
'Oh, deary me!'
The King sobbed, 'Oh, deary me
And went back to bed.
'Nobody,'
He whimpered,
'Could call me
A fussy man;
I *only* want
A little bit
Of butter for
My bread!'

The Queen said,
'There, there!'
And went to
The Dairymaid.
The Dairymaid.
Said, 'There, there!'
And went to the shed.
The cow said,
'There, there!
I didn't really
Mean it;
Here's milk for his porringer
And butter for his bread.'

The Queen took
The butter
And brought it to
His Majesty;
The King said,
'Butter, eh?'
And bounced out of bed.
'Nobody, he said,
As he kissed her
Tenderly,
'Nobody,' he said,
As he slid down
The banisters,
'Nobody,
My darling,
Could call me
A fussy man —
BUT
I do like a little bit of butter to my bread!'

A. A. Milne

71

Queen Nefertiti

Spin a coin, spin a coin,
   All fall down;
Queen Nefertiti
   Stalks through the town.

Over the pavements
   Her feet go clack.
Her legs are as tall
   As a chimney stack;

Her fingers flicker
   Like snakes in the air,
The walls split open
   At her green-eyed stare;

Her voice is thin
   As the ghosts of bees;
She will crumble your bones
   She will make your blood freeze.

Spin a coin, spin a coin,
   All fall down,
Queen Nefertiti
   Stalks through the town.

Anon.

Not a Very Cheerful Song, I'm Afraid

There was a gloomy lady,
With a gloomy duck and a gloomy drake,
And they all three wandered gloomily,
Beside a gloomy lake,
On a gloomy, gloomy, gloomy, gloomy, gloomy, gloomy day.

Now underneath that gloomy lake
The gloomy lady's gone.
But the gloomy duck and the gloomy drake
Swim on and on and on,
On a gloomy, gloomy, gloomy, gloomy, gloomy, gloomy day.

Adrian Mitchell

Johnnie Crack and Flossie Snail

John Crack and Flossie Snail
Kept their baby in a milking pail
Flossie Snail and Johnnie Crack
One would pull it out and one would put it back

O it's my turn now said Flossie Snail
To take the baby from the milking pail
And it's my turn now said Johnnie Crack
To smack it on the head and put it back

Johnnie Crack and Flossie Snail
Kept their baby in a milking pail
One would put it back and one would pull it out
And all it had to drink was ale and stout
For Johnnie Crack and Flossie Snail
Always used to say that stout and ale
Was *good* for a baby in a milking pail.

Dylan Thomas

Old Mrs Lazibones

Old Mrs Lazibones
And her dirty daughter
Never used soap
And never used water.
    Higgledy piggledy cowpat
    What d'you think of that?

Daisies from their fingernails,
Birds' nests in their hair-O
Dandelions from their ears, –
What a dirty pair-O!
    Higgledy piggedly cowpat
    What d'you think of that?

Came a prince who sought a bride,
Riding past their doorstep,
Quick, said Mrs Lazibones.
Girl, under the watertap.
    Higgledy piggledy cowpat
    What d'you think of that?

Washed her up and washed her down,
Then she washed her sideways,
But the prince was far, far away,
He'd ridden off on the highways.
    Higgledy piggledy cowpat
    What d'you think of that?

Gerda Mayer

Mrs Lorris, Who Died of Being Clean

Mrs Lorris was a fusser
always asking, 'Is it clean?'
boiled her knives and forks twice daily,
vacuumed the village green.
When she took the bus to market
she spread a towel on the seat,
washed her hands in disinfectant
before sitting down to eat.
She never ate raw food like lettuce –
'Full of germs,' she used to say
and by her strenuous housecleaning
hoped to keep the germs away.
Always at it, night and morning,
with the scrubbing-brush and soap –
still she wasn't really certain
so she bought a microscope.
Horrid horror, in the eyepiece
microbes swarmed on every side:
too much for Mrs Lorris, who
pegged her nostrils up and died.

Barbara Giles

Greedyguts

I sat in the café and sipped at a Coke.
There sat down beside me a WHOPPING great bloke
Who sighed as he elbowed me into the wall:
'Your trouble, my boy, is your belly's too small!
Your bottom's too thin! Take a lesson from me:
I may not be nice, but I'm GREAT, you'll agree,
And I've lasted a lifetime by playing this hunch:
The bigger the breakfast, the larger the lunch!

The larger the lunch, then the huger the supper.
The deeper the teapot, the vaster the cupper.
The fatter the sausage, the fuller the tea.
The MORE on the table, the BETTER FOR ME!'

His elbows moved in and his elbows moved out,
His belly grew bigger, chins wobbled about,
As forkful by forkful and plate after plate,
He ate and he ate and he ate and he ATE!

I hardly could breathe, I squashed out of shape,
So under the table I made my escape.

'Aha!' he rejoiced, 'when it's put to the test,
The fellow who's fattest will come off the best!
Remember, my boy, when it comes to the crunch:
The bigger the breakfast, the larger the lunch!

The larger the lunch, then the huger the supper.
The deeper the teapot, the vaster the cupper.
The fatter the sausage, the fuller the tea.
The MORE on the table, the BETTER FOR ME!'

A lady came by who was scrubbing the floor
With a mop and a bucket. To even the score,
I lifted that bucket of water and said,
As I poured the whole lot of it over his head:

'*I've* found all my life, it's a pretty sure bet:
The FULLER the bucket, the WETTER YOU GET!'

Kit Wright

Ned

It's a singular thing that Ned
Can't be got out of bed.
    When the sun comes round
    He is sleeping sound
With blankets over his head.
    They tell him to shunt,
    And he gives a grunt,
And burrows a little deeper –
    He's a trial to them
    At eight a.m.,
When Ned is a non-stop sleeper.
    Oh, the snuggly bits
    Where the pillow fits
    Into his cheeks and neck!
    Oh, the beautiful heat
    Stored under the sheet
Which the breakfast-bell will wreck!
Oo, the noozly-oozly feel
He feels from head to heel,
    When to get out of bed
    Is worse to Ned
Than missing his morning meal!
                But
It's a singular thing that Ned,
    After the sun is dead
    And the moon's come round,
    Is not to be found,
    And can't be got into bed!

Eleanor Farjeon

Noah

When Noah sailed the wet and blue
He took in the animals two and two.

When the animals made too much of a din,
There was water without, but wine within.

The bulls they bellowed, the ducks made quack,
The geese went honk, but Noah lay back.

The cuckoo went cuckoo, the owls went hoot,
Mrs Noah said much, but Noah was mute.

And Noah lay in his hammock all day,
While the ark continued its watery way.

The dogs cried woof, but the cat gave a purr,
Because Ham, Shem and Japheth, were stroking his fur.

Ham slipped up but his brothers were nifty,
And Noah lived on to nine hundred and fifty.

Gerda Mayer

Noah

Noah was an Admiral;
Never a one but he
Sailed for forty days and nights
With wife and children three
On such a mighty sea.

Under his tempest-battered deck
This Admiral had a zoo;
And all the creatures in the world,
He kept them, two by two —
Ant, hippo, kangaroo,

And every other beast beside,
Of every mould and make.
When tempests howled and thunder growled
How they did cower and quake
To feel the vessel shake!

But Noah was a Carpenter
Had made his ship so sound
That not a soul of crew or zoo
In all that time was drowned
Before they reached dry ground.

So Admiral, Keeper, Carpenter —
Now should *you* put to sea
In such a flood, it would be good
If one of these you be,
But better still — all three!

James Reeves

Tall Story for Fred Dibnah

Jack Steeplejack,
A joker,
Mended chimbleys
Up North.
One day
The lad
Who worked
For him
Asked for
His cards.
'Why?' said
Jack, sad.
'Going down
Pit,' lad said.
'But why?'
'I'm tired
Of heights,'
Lad said.
Next day
Jack had
New lad
To do chores,
Odd jobs,
Brew tea,
Hod-jobs,
Mix mortar,
Clean bricks,
Push trolley,
Hold brolly
When it rained
Cats and dogs.
Twelve o'clock

Right up
Stack-top

Jack said
To lad,
'Time now
For snap.'
'Oh, I didn't
Bring ought,'
Said lad,
Tummy
Rumbling.
'No sweat,'
Jack said.
'You fetch it.
Chish 'n' fips.
Here's a quid.
Be quick.
Mind, now, how
You go.'
Lad clumb
Down ladder,
Five hundred
And eighty-
Five rungs.

Walked a
Bit unsteady
To fish shop,
Queued for
Cod 'n' chips.
Climbed up
Five hundred
And eighty-
Five rungs.

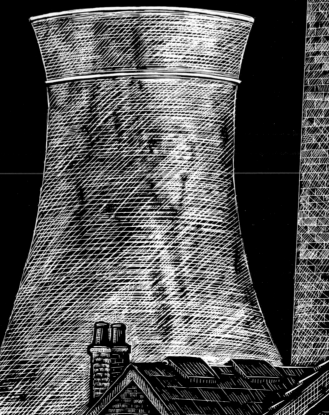

'Thanks, lad,'
Jack said.
'You're a brick.'
Jack, quick,
Unwrapped
Newspaper.
Then said,
'Where's
Vinegar?'
'Vinegar?'
Said lad,
Out of
Breath.
'Aye, you've
Been had,
My lad.

They missed
Vinegar.
Eh?
Whoever
Heard of
Chips
Without?
Eh?'
Lad said,
'Well, I'll
Go and
Get it.'
'No, lad,'
Said Jack,
'It's just
My joke.
Eat up
Your snap.

Eh, but
Didn't
You put
Any salt
On, eh?
Better go
And get
Some salt.'

That lad
Worked for
Jack for
Twenty years,
Man and boy.
Never did
Know when
Jack was
Joking.

Geoffrey Summerfield

The Pobble who has no Toes

The Pobble who has no toes
    Had once as many as we;
When they said, 'Some day you may lose them all,'
    He replied, – 'Fish fiddle de-dee!'
And his Aunt Jobiska made him drink,
Lavender water tinged with pink,
For she said, 'The World in general knows
There's nothing so good for a Pobble's toes!'

The Pobble who has no toes,
    Swam across the Bristol Channel;
But before he set out he wrapped his nose
    In a piece of scarlet flannel.
For his Aunt Jobiska said, 'No harm
Can come to his toes if his nose is warm;
And it's perfectly known that a Pobble's toes
Are safe, – provided he minds his nose.'

The Pobble swam fast and well,
    And when boats or ships came near him
He tinkledy-binkledy-winkled a bell,
    So that all the world could hear him.
And all the Sailors and Admirals cried,
When they saw him nearing the further side, –
'He has gone to fish, for his Aunt Jobiska's
Runcible Cat with crimson whiskers!'

But before he touched the shore,
    The shore of the Bristol Channel,
A sea-green Porpoise carried away
    His wrapper of scarlet flannel.
And when he came to observe his feet,
Formerly garnished with toes so neat,
His face at once became forlorn
On perceiving that all his toes were gone!

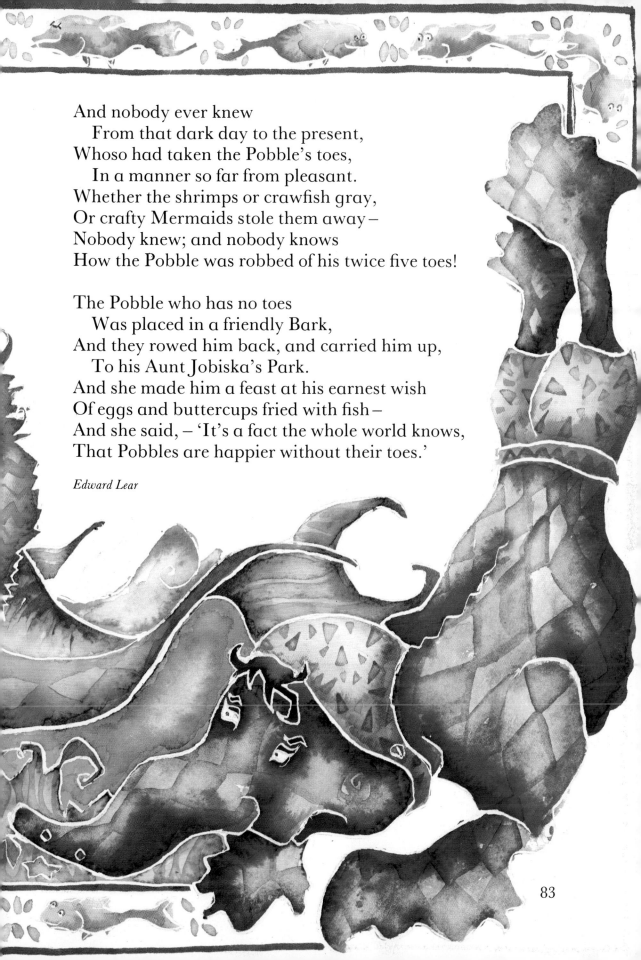

And nobody ever knew
   From that dark day to the present,
Whoso had taken the Pobble's toes,
   In a manner so far from pleasant.
Whether the shrimps or crawfish gray,
Or crafty Mermaids stole them away –
Nobody knew; and nobody knows
How the Pobble was robbed of his twice five toes!

The Pobble who has no toes
   Was placed in a friendly Bark,
And they rowed him back, and carried him up,
   To his Aunt Jobiska's Park.
And she made him a feast at his earnest wish
Of eggs and buttercups fried with fish –
And she said, – 'It's a fact the whole world knows,
That Pobbles are happier without their toes.'

Edward Lear

The Quangle Wangle's Hat

On top of the Crumpetty Tree
  The Quangle Wangle sat,
But his face you could not see,
  On account of his Beaver Hat.
For his Hat was a hundred and two feet wide,
With ribbons and bibbons on every side,
And bells, and buttons, and loops, and lace,
So that nobody ever could see the face
    Of the Quangle Wangle Quee.

The Quangle Wangle said
  To himself on the Crumpetty Tree,
'Jam; and jelly; and bread;
  Are the best of food for me!
But the longer I live on this Crumpetty Tree,
The plainer than ever it seems to me
That very few people come this way,
And that life on the whole is far from gay!'
    Said the Quangle Wangle Quee.

But there came to the Crumpetty Tree
  Mr and Mrs Canary;
And they said, 'Did ever you see
  Any spot so charmingly airy?
May we build a nest on your lovely Hat?
Mr Quangle Wangle, grant us that!
O please let us come and build a nest
Of whatever material suits you best,
    Mr Quangle Wangle Quee!'

And besides, to the Crumpetty Tree
   Came the Stork, the Duck, and the Owl;
The Snail and the Bumble-Bee,
   The Frog, and the Fimble Fowl
(The Fimble Fowl with a corkscrew leg);
And all of them said, 'We humbly beg,
We may build our homes on your lovely Hat,
Mr Quangle Wangle, grant us that!
    Mr Quangle Wangle Quee!'

And the Golden Grouse came there,
   And the Pobble who hast no toes,
And the small Olympian Bear
   And the Dong with a luminous nose.
And the Blue Baboon, who played the flute,
And the Orient Calf from the Land of Tute,
And the Attery Squash and the Bisky Bat,
All came and built on the lovely Hat
   Of the Quangle Wangle Quee.

And the Quangle Wangle said
   To himself on the Crumpetty Tree,
'When all these creatures move
   What a wonderful noise there'll be!'
And at night by the light of the Mulberry Moon
They danced to the Flute of the Blue Baboon
On the broad green leaves of the Crumpetty Tree,
And all were as happy as happy could be,
   With the Quangle Wangle Quee.

Edward Lear

Hurt no Living Thing

Hurt no living thing;
Ladybird, nor butterfly,
Nor moth with dusty wing,
Nor cricket chirping cheerily,
Nor grasshopper so light of leap,
Nor dancing gnat, nor beetle fat,
Nor harmless worms that creep.

Christina Rossetti

A Fox Came into my Garden

A fox came into my garden.
'What do you want from me?'
'Heigh-ho, Johnnie-boy,
A chicken for my tea.'

'Oh no, you beggar, and never, you thief,
My chicken you must leave,
That she may run and she may fly
From now to Christmas Eve.'

'What are you eating, Johnnie-boy,
Between two slices of bread?'
'I'm eating a piece of chicken-breast
And it's honey-sweet,' I said.

'Heigh-ho, you diddling man,
I thought that was what I could smell.
What, some for you and none for me?
Give us a piece as well!'

Charles Causley

The Answers

'When did the world begin and how?'
I asked a lamb,
   a goat,
      a cow.
'What is it all about and why?'
I asked a hog as he went by.
'Where will the whole thing end and when?'
I asked a duck,
   a goose
      a hen:
And I copied all the answers too,
A quack
   a honk
      an oink
         a moo.

Robert Clairmont

Animals' Houses

Of animals' houses
  Two sorts are found –
Those which are square ones
  And those which are round.

Square is a hen-house,
  A kennel, a sty:
Cows have square houses
  And so have I.

A snail's shell is curly,
  A bird's nest round;
Rabbits have twisty burrows
  Underground.

But the fish in the bowl
  And the fish at sea –
Their houses are round
  As a house can be.

James Reeves

Heads or Tails?

Dave Dirt's dog is a horrible hound,
  A hideous sight to see.
When Dave first brought it home from the pound,
We couldn't be certain which way round
  The thing was supposed to be!

Somebody said, 'If that's its *head*,
  It's *far* the ugliest dog in town.'
Somebody said, 'The darned thing's *dead*!'
  'Don't be silly, it's *upside-down*!'
'It's *inside-out*!' 'It's a sort of *plant*!'
'It's wearing *clothes*!' 'It's Dave Dirt's *aunt*!'
  'It's a sort of *dressing-gown*!'

Each expert had his own idea
  Of what it was meant to be
But everybody was far from clear –
  And yet . . . we *did* agree
That Dave Dirt's dog was a horrible hound
  And a hideous sight to see!

It *loves Dave Dirt*. It follows him round
  Through rain and sun and snow.
When set in motion, it looks far *worse*,
And nobody knows if it's in reverse
  Or the way it's supposed to go!

Kit Wright

My Dog

Have you seen a little dog anywhere about?
A raggy dog, a shaggy dog,
    who's always looking out
For some fresh mischief which he thinks
    he really ought to do.
He's very likely at this minute
    biting someone's shoe.

If you see that little dog,
    his tail up in the air,
A whirly tail, a curly tail,
    a dog who doesn't care
For any other dog he meets,
    not even for himself,
Then hide your mats, and put your meat
    upon the top-most shelf.

If you see that little dog, barking at the cars,
A raggy dog, a shaggy dog,
    with eyes like twinkling stars,
Just let me know, for though he's bad,
    as bad as bad can be,
I wouldn't change that dog for all
    the treasures of the sea.

Emily Lewis

Our Hamster's Life

Our hamster's life:
there's not much
to it,
not much
to it.

He presses his pink nose
to the door of his cage
and decides for the fifty six
millionth time
that he can't get
through it.

Our hamster's life;
there's not much
to it,
not much
to it.

It's about the most boring
life in the world,
if he only
knew it.
He sleeps and he drinks
    and he eats.
He eats and he drinks
    and he sleeps.

He slinks and he dreeps.
He eats.

This process
he repeats.

Our hamster's life:
there's not much
to it,
not much
to it.

You'd think it would drive
    him bonkers,
going round and round
    on his wheel.
It's certainly driving
    me bonkers,

watching him
do it.

But he may be thinking:
'That boy's life,
there's not much
to it,
not much
to it:

watching a hamster go round
    on a wheel.
It's driving me bonkers if he
    only knew it,

watching him
watching me
do it.'

Kit Wright

My New Rabbit

We brought him home, I was so pleased,
   We made a rabbit-hutch,
I give him oats, I talk to him,
   I love him very much.

Now when I talk to Rover dog,
   He answers me 'Bow-wow!'
And when I speak to Pussy-cat,
   She purrs and says 'Mee-ow!'

But Bunny never says a word,
   Just twinkles with his nose,
And what that rabbit thinks about,
   Why! no one ever knows.

My Mother says the fairies must
   Have put on him a spell,
They told him all their secrets, then
   They whispered, 'Pray don't tell.'

So Bunny sits there looking wise,
   And twinkling with his nose,
And never, never, never tells
   A single thing he knows.

Elizabeth Gould

This and That

Two cats together
In bee-heavy weather
After the August day
In smug contentment lay
By the garden shed
In the flower bed
Yawning out the hours
In the shade of the flowers
And passed the time away,
Between stretching and washing and sleeping,
Talking over the day.

'Climbed a tree.'
'Aaaah.'
'Terrorized sparrows.'
'Mmmmh.'
'Was chased.'
'Aaaah.'
'Fawned somewhat!'
'Mmmmh.'
'Washed, this and that,'
Said the first cat.

And they passed the time away
Between stretching and washing and sleeping
Talking over the day.

'Gazed out of parlour window.'
'Aaaah.'
'Pursued blue bottles.'
'Mmmmh.'
'Clawed curtains.'
'Aaaah.'
'Was cuffed.'
'Mmmmh.'

'Washed, this and that.'
Said the other cat.

And they passed the time away
Between stretching and washing and sleeping
Talking over the day.

'Scratched to be let in.'
'Aaaah.'
'Patrolled the house.'
'Mmmmh.'
'Scratched to go out.'
'Aaaah.'
'Was booted.'
'Mmmmh.'
'Washed, this and that.'
Said the first cat.

And they passed the time away
Between stretching and washing and sleeping
Talking over the day.

'Lapped cream elegantly.'
'Aaaah.'
'Disdained dinner.'
'Mmmmh.'
'Borrowed a little salmon.'
'Aaaah.'
'Was tormented.'
'Mmmmh.'
'Washed, this and that.'
Said the other cat.

And they passed the time away
Between stretching and washing and sleeping
Talking over the day.

Gareth Owen

My Uncle Paul of Pimlico

My Uncle Paul of Pimlico
Has seven cats as white as snow,
Who sit at his enormous feet
And watch him, as a special treat,
Play the piano upside-down,
In his delightful dressing-gown;
The firelight leaps, the parlour glows,
And, while the music ebbs and flows,
They smile (while purring the refrains),
At little thoughts that cross their brains.

Mervyn Peake

Montague Michael

Montague Michael
You're much too fat,
You wicked old, wily old,
Well-fed cat.

All night you sleep
On a cushion of silk,
And twice a day
I bring you milk.

And once in a while,
When you catch a mouse,
You're the proudest person
In all the house.

But spoilt as you are,
I tell you sir,
This dolly is mine
And you can't have her!

Anon.

94

Cats

Cats sleep
  Anywhere,
    Any table,
      Any chair,
        Top of piano,
          Window-ledge,
           In the middle,
            On the edge,
             Open drawer,
              Empty shoe,
               Anybody's
                Lap will do,
                 Fitted in a
                  Cardboard box,
                 In the cupboard
                 With your frocks –
               Anywhere!
                   They don't care!
                Cats sleep
                 Anywhere.

Eleanor Farjeon

Alley Cat

A bit of jungle in the street
He goes on velvet toes,
And slinking through the shadows, stalks
Imaginary foes.

Esther Valck Georges

The Greater Cats

The greater cats with golden eyes
Stare out between the bars.
Deserts are there, and different skies,
And night with different stars.

Victoria Sackville-West

The Tiger

The tiger behind the bars of his cage growls,
The tiger behind the bars of his cage snarls,
The tiger behind the bars of his cage roars.
        Then he thinks.
It would be nice not to be behind bars all
        The time
        Because they spoil my view
        I wish I were wild, not on show.
But if I were wild, hunters might shoot me,
But if I were wild, food might poison me,
But if I were wild, water might drown me.
        Then he stops thinking
        And . . .
The tiger behind the bars of his cage growls,
The tiger behind the bars of his cage snarls,
The tiger behind the bars of his cage roars.

Peter Niblett

A Tiger in the Zoo

He stalks in his vivid stripes
The few steps of his cage,
On pads of velvet quiet,
In his quiet rage.

He should be lurking in shadow,
Sliding through long grass
Near the water hole
Where plump deer pass.

He should be snarling around houses
At the jungle's edge,
Baring his white fangs, his claws,
Terrorising the village!

But he's locked in a concrete cell,
His strength behind bars,
Stalking the length of his cage,
Ignoring visitors.

He hears the last voice at night,
The patrolling cars,
And stares with his brilliant eyes
At the brilliant stars.

Leslie Norris

The Owl and the Pussy-cat

The Owl and the Pussy-cat went to sea
    In a beautiful pea-green boat,
They took some honey, and plenty of money,
    Wrapped up in a five-pound note.
The Owl looked up to the stars above,
    And sang to a small guitar,
'O lovely Pussy! O Pussy, my love,
    What a beautiful Pussy you are,
        You are,
        You are!
    What a beautiful Pussy you are!'

Pussy said to the Owl, 'You elegant fowl!
    How charmingly sweet you sing!
O let us be married! too long we have tarried:
    But what shall we do for a ring?'
They sailed away, for a year and a day,
    To the land where the Bong-tree grows,
And there in a wood a Piggy-wig stood
    With a ring at the end of his nose,
        His nose,
        His nose,
    With a ring at the end of his nose.

'Dear Pig, are you willing to sell for one shilling
    Your ring?' Said the Piggy, 'I will.'
So they took it away, and were married next day
    By the turkey who lives on the hill.
They dined on mince, and slices of quince,
    Which they ate with a runcible spoon;
And hand in hand, on the edge of the sand,
    They danced by the light of the moon,
        The moon,
        The moon,
    They danced by the light of the moon.

Edward Lear

The Mouse in the Wainscot

Hush, Suzanne!
Don't lift your cup.
That breath you heard
Is a mouse getting up.

As the mist that steams
From your milk as you sup,
So soft is the sound
Of a mouse getting up.

There! did you hear
His feet pitter-patter,
Lighter than tipping
Of beads in a platter,

And then like a shower
On the window pane
The little feet scampering
Back again?

O falling of feather!
O drift of a leaf!
The mouse in the wainscot
Is dropping asleep.

Ian Serraillier

Wishes

Said the first little chicken
With a queer little squirm,
'I wish I could find
A fat little worm.'

Said the second little chicken
With an odd little shrug,
'I wish I could find
A fat little slug.'

Said the third little chicken
With a sharp little squeal,
'I wish I could find
Some nice yellow meal!'

Said the fourth little chicken
With a small sigh of grief,
'I wish I could find
A little green leaf.'

Said the fifth little chicken
With a faint little moan,
'I wish I could find
A small gravel stone.'

'Now see here,' said their mother
From the green garden patch.
'If you want any breakfast,
Just come here and
        SCRATCH!'

Anon.

Duck's Ditty

All along the backwater,
   Through the rushes tall,
Ducks are a-dabbling,
   Up tails all!

Ducks' tails, drakes' tails,
   Yellow feet a-quiver,
Yellow bills all out of sight
   Busy in the river!

Slushy green undergrowth
   Where the roach swim,
Here we keep our larder
   Cool and full and dim!

Every one for what he likes!
   We like to be
Heads down, tails up,
   Dabbling free!

High in the blue above
   Swifts whirl and call –
We are down a-dabbling,
   Up tails all!

Kenneth Grahame

Cage Bird and Sky Bird

Cage Bird swung
From an apple tree
And his cage was of silver
And ivory.
   Nobody can be so happy, so happy;
      Sang the Cage Bird.

Sky Bird Sang
From a cloudless sky,
And his wings were wide
And bright his eye.
   Nobody can be as happy as I;
      Sang the Sky Bird.

That wild song
To the garden fell
And Cage Bird heard it, in
His silver cell.
   Are you free then, are you truly free?
      Cried the Cage Bird.

Sky Bird flew
In the trail of the sun
And swiftly he soared, away
From the garden.
   Sadly sang Cage Bird, when the day was done;
      Sang the Cage Bird.

Leslie Norris

Owl of the Greenwood

'Owl
Who?
Who are you?
Who?'

      'I am owl,
      night's eyes,
      wise beyond understanding.'

'Who?
Who are you?
Who?'

      'I am owl,
      shadow of shadows,
      owner of forests,
      beautiful beyond comprehension.'

'Who?
Who are you?
Who?'

      'I am owl,
      plucker of moonbeams;
      owl, most mysterious.
      Beware.'

Patricia Hubbell

Yesterday the House was full of Flies

One went spinning down the plughole,
Clinging to a tea-leaf.
Two pestered the dog. He snapped, and caught them.
He was as surprised as the flies.
Three sat all day on a fruit-loaf,
Disguised as currants.
Four zizzed in a spider's web,
Until the spider woke up.
Five chased each other round a lamp-shade
Until they were giddy.
Six padded up and down the windows,
And still can't fathom glass.
Seven sat on the warm electric kettle,
Until I switched it on.
Eight stuck to reading all about glue
In the fly-paper.
Nine played on a broken fly-swatter,
Laughing themselves silly.
Ten walked all over the mirror,
Admiring their stomachs.
Eleven pestered each other, trying to bark,
Doing an imitation of a dog-fight.
Twelve went supersonic into the window,
Knocked themselves out.
And hundreds just sat for hours,
Twiddling their legs.
I folded a paper, killed one,
And a thousand more came to its funeral.

Geoffrey Summerfield

The Fly

How large unto the tiny fly
Must little things appear! –
A rosebud like a feather bed,
Its prickle like a spear;

A dewdrop like a looking-glass,
A hair like golden wire;
The smallest grain of mustard seed
As fierce as coals of fire;

A loaf of bread, a lofty hill;
A wasp, a cruel leopard;
And specks of salt as bright to see
As lambkins to a shepherd.

Walter de la Mare

The Grasshopper and the Bird

The grasshopper said
To the bird in the tree
  Zik-a-zik zik-a-zik
As polite as could be—
  Zik-a-zik zik-a-zik—
Which he meant for to say
In his grasshopper way
For the time of the year
'Twas a vairy warm day—
  Zik-a-zik zik-a-zik—
What a very warm day!

  Tee-oo-ee tee-oo-ee
Said the bird in the tree,
  Tee-oo-ee tee-oo-ee
As polite as could be;
That's as much as to say—
  Tee-oo-ee tee-oo-ee
That I can't quite agree!
So he upped with his wings—
  Tee-oo-ee tee-oo-ee
And he flew from the tree.

So the grasshopper hopped
Four hops and away
  Snick!
    Click!
      Flick!
        Slick!—
Four hops and away
To the edge of the hay
  Zik-a-zik zik-a-zik
For the rest of the day.

  Zik-a-zik zik-a-zik
  Tee-oo-ee tee-oo-ee
The bird and the grasshopper
Can't quite agree.

James Reeves

The Bee's Last Journey to the Rose

I came first through the warm grass
Humming with spring,
And now swim through the evening's
Soft sunlight gone cold.
I'm old in this green ocean,
Going a final time to the rose.

North Wind, until I reach it,
Keep your icy breath away
That changes pollen into dust.
Let me be drunk on this scent a final time.
Then blow if you must.

Brian Patten

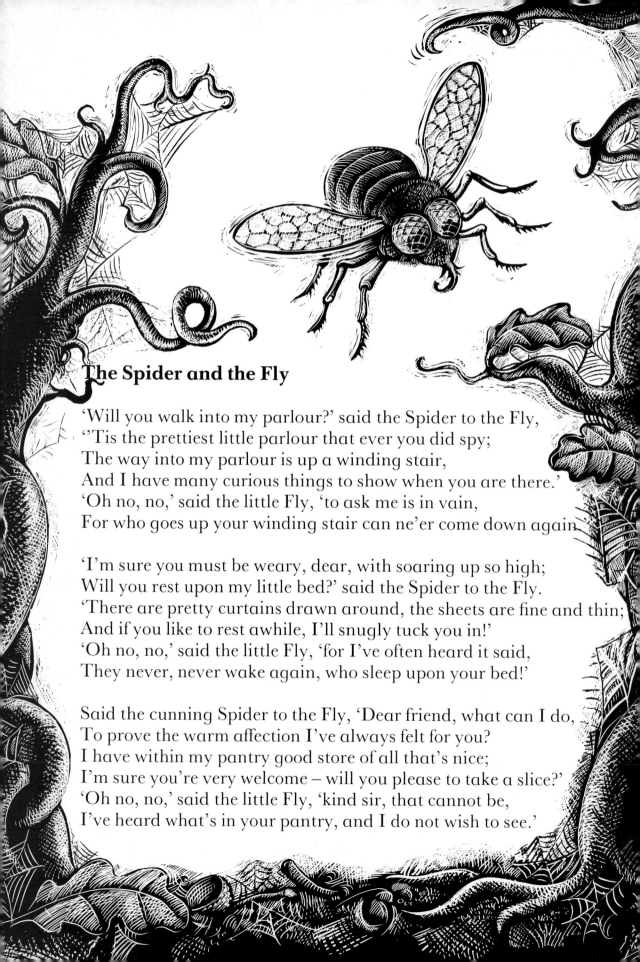

The Spider and the Fly

'Will you walk into my parlour?' said the Spider to the Fly,
''Tis the prettiest little parlour that ever you did spy;
The way into my parlour is up a winding stair,
And I have many curious things to show when you are there.'
'Oh no, no,' said the little Fly, 'to ask me is in vain,
For who goes up your winding stair can ne'er come down again.'

'I'm sure you must be weary, dear, with soaring up so high;
Will you rest upon my little bed?' said the Spider to the Fly.
'There are pretty curtains drawn around, the sheets are fine and thin;
And if you like to rest awhile, I'll snugly tuck you in!'
'Oh no, no,' said the little Fly, 'for I've often heard it said,
They never, never wake again, who sleep upon your bed!'

Said the cunning Spider to the Fly, 'Dear friend, what can I do,
To prove the warm affection I've always felt for you?
I have within my pantry good store of all that's nice;
I'm sure you're very welcome – will you please to take a slice?'
'Oh no, no,' said the little Fly, 'kind sir, that cannot be,
I've heard what's in your pantry, and I do not wish to see.'

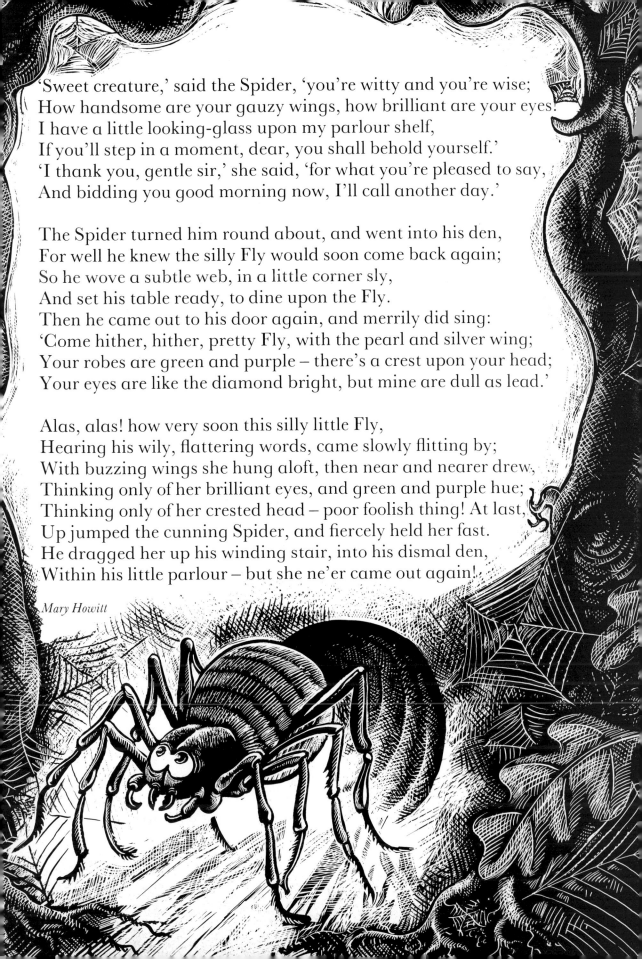

'Sweet creature,' said the Spider, 'you're witty and you're wise;
How handsome are your gauzy wings, how brilliant are your eyes!
I have a little looking-glass upon my parlour shelf,
If you'll step in a moment, dear, you shall behold yourself.'
'I thank you, gentle sir,' she said, 'for what you're pleased to say,
And bidding you good morning now, I'll call another day.'

The Spider turned him round about, and went into his den,
For well he knew the silly Fly would soon come back again;
So he wove a subtle web, in a little corner sly,
And set his table ready, to dine upon the Fly.
Then he came out to his door again, and merrily did sing:
'Come hither, hither, pretty Fly, with the pearl and silver wing;
Your robes are green and purple – there's a crest upon your head;
Your eyes are like the diamond bright, but mine are dull as lead.'

Alas, alas! how very soon this silly little Fly,
Hearing his wily, flattering words, came slowly flitting by;
With buzzing wings she hung aloft, then near and nearer drew,
Thinking only of her brilliant eyes, and green and purple hue;
Thinking only of her crested head – poor foolish thing! At last,
Up jumped the cunning Spider, and fiercely held her fast.
He dragged her up his winding stair, into his dismal den,
Within his little parlour – but she ne'er came out again!

Mary Howitt

The Walrus and the Carpenter

The sun was shining on the sea,
  Shining with all his might:
He did his very best to make
  The billows smooth and bright —
And this was odd, because it was
  The middle of the night.

The moon was shining sulkily,
  Because she thought the sun
Had got no business to be there
  After the day was done —
'It's very rude of him,' she said,
  'To come and spoil the fun!'

The sea was wet as wet could be,
  The sands were dry as dry.
You could not see a cloud, because
  No cloud was in the sky:
No birds were flying overhead —
  There were no birds to fly.

The Walrus and the Carpenter
  Were walking close at hand;
They wept like anything to see
  Such quantities of sand:
'If this were only cleared away,'
  They said, 'It *would* be grand!'

'If seven maids with seven mops
  Swept it for half a year,
Do you suppose,' the Walrus said,
  'That they could get it clear?'
'I doubt it,' said the Carpenter,
  And shed a bitter tear.

'O Oysters, come and walk with us!'
　The Walrus did beseech.
'A pleasant walk, a pleasant talk,
　Along the briny beach;
We cannot do with more than four,
　To give a hand to each.'

The eldest Oyster looked at him,
　But never a word he said:
The eldest Oyster winked his eye,
　And shook his heavy head—
Meaning to say he did not choose
　To leave the oyster-bed.

But four young Oysters hurried up,
　All eager for the treat:
Their coats were brushed, their faces washed,
　Their shoes were clean and neat—
And this was odd, because, you know,
　They hadn't any feet.

Four other Oysters followed them,
    And yet another four;
And thick and fast they came at last,
    And more, and more, and more—
All hopping through the frothy waves,
    And scrambling to the shore.

The Walrus and the Carpenter
    Walked on a mile or so,
And then they rested on a rock
    Conveniently low:
And all the little Oysters stood
    And waited in a row.

'The time has come,' the Walrus said,
    'To talk of many things:
Of shoes—and ships—and sealing-wax—
    Of cabbages—and kings—
And why the sea is boiling hot—
    And whether pigs have wings.'

'But wait a bit,' the Oysters cried,
    'Before we have our chat;
For some of us are out of breath,
    And all of us are fat!'
'No hurry!' said the Carpenter.
    They thanked him much for that.

'A loaf of bread,' the Walrus said,
    'Is what we chiefly need:
Pepper and vinegar besides
    Are very good indeed—
Now if you're ready, Oysters dear,
    We can begin to feed.'

'But not on us!' the Oysters cried,
    Turning a little blue.
'After such kindness, that would be
    A dismal thing to do!'
'The night is fine,' the Walrus said,
    'Do you admire the view?

'It was so kind of you to come!
    And you are very nice!'
The Carpenter said nothing but
    'Cut us another slice:
I wish you were not quite so deaf—
    I've had to ask you twice!'

'It seems a shame,' the Walrus said,
    'To play them such a trick,
After we've brought them out so far,
    And made them trot so quick!'
The Carpenter said nothing but
    'The butter's spread too thick!'

'I weep for you,' the Walrus said:
    'I deeply sympathize.'
With sobs and tears he sorted out
    Those of the largest size,
Holding his pocket-handkerchief
    Before his streaming eyes.

'O Oysters,' said the Carpenter,
    'You've had a pleasant run!
Shall we be trotting home again?'
    But answer came there none—
And this was scarcely odd, because
    They'd eaten every one.

Lewis Carroll

Goldfish

Our headteacher has a golden
tooth. It's real gold. I asked him,
so I know. 'Are you very rich?'
I asked. He laughed and, like a fish

rising up its pond to feed, the tooth
flashed gold and surfaced in his mouth.
I watched its glint with fascination
and hardly heard his conversation.

I saw it had a healthy shine
of glittering beams pulled from the sun.
It nibbled in between the fronds
and waterweed of lips and tongue,

then dived again into the gloom.
Now when he comes in our classroom
I play our newest game of guile –
feeding the goldfish with a smile.

Barrie Wade

It makes a Change

There's nothing makes a Greenland Whale
Feel half so high-and-mighty,
As sitting on a mantelpiece
In Aunty Mabel's nighty.

It makes a change from Freezing Seas,
(Of which a Whale can tire),
To warm his weary tail at ease
Before an English fire.

For this delight he leaves the sea,
(Unknown to Aunty Mabel),
Returning only when the dawn
Lights up the Breakfast Table.

Mervyn Peake

The Blind Men and the Elephant

It was six men of Hindostan,
  To learning much inclined,
Who went to see the elephant,
  (Though all of them were blind):
That each by observation
  Might satisfy his mind.

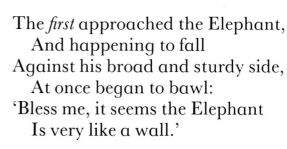

The *first* approached the Elephant,
  And happening to fall
Against his broad and sturdy side,
  At once began to bawl:
'Bless me, it seems the Elephant
  Is very like a wall.'

The *second*, feeling of his tusk,
  Cried, 'Ho! what have we here
So very round and smooth and sharp?
  To me 'tis mighty clear
This wonder of an Elephant
  Is very like a spear.'

The *third* approached the animal,
  And happening to take
The squirming trunk within his hands,
  Then boldly up and spake:
'I see,' quoth he, 'the Elephant
  Is very like a snake.'

The *fourth* stretched out his eager hand
    And felt about the knee,
'What most this mighty beast is like
    Is mighty plain,' quoth he;
''Tis clear enough the Elephant
    Is very like a tree.'

The *fifth* who chanced to touch the ear
    Said, 'Even the blindest man
Can tell what this resembles most;
    Deny the fact who can,
This marvel of an Elephant
    Is very like a fan.'

The *sixth* no sooner had begun
    About the beast to grope,
Than, seizing the swinging tail
    That fell within his scope,
'I see,' cried he, 'the Elephant
    Is very like a rope.'

And so these men of Hindostan
    Disputed loud and long,
Each in his own opinion
    Exceeding stiff and strong,
Though *each* was *partly* in the right
And all were in the wrong.

John Godfrey Saxe

The Song of the Camel

'Canary-birds feed on sugar and seed,
  Parrots have crackers to crunch;
And, as for the poodles, they tell me the noodles
  Have chickens and cream for their lunch.
    But there's never a question
    About MY digestion –
  ANYTHING does for me!

'Cats, you're aware, can repose in a chair,
  Chickens can roost upon rails;
Puppies are able to sleep in a stable,
  And oysters can slumber in pails.
    But no one supposes
    A poor Camel dozes –
  ANY PLACE does for me!

'Lambs are enclosed where it's never exposed,
  Coops are constructed for hens;
Kittens are treated to houses well heated,
  And pigs are protected by pens.
    But a Camel comes handy
    Wherever it's sandy –
  ANYWHERE does for me!

'People would laugh if you rode a giraffe,
  Or mounted the back of an ox;
It's nobody's habit to ride on a rabbit,
  Or try to bestraddle a fox.
    But as for a Camel, he's
    Ridden by families –
  ANY LOAD does for me!

'A snake is as round as a hole in the ground,
　And weasels are wavy and sleek;
And no alligator could ever be straighter
　Than lizards that live in a creek.
　　But a Camel's all lumpy
　　And bumpy and humpy –
　ANY SHAPE does for me!'

Charles Edward Carryl

119

The Tale of Custard the Dragon

Belinda lived in a little white house,
With a little black kitten and a little grey mouse,
And a little yellow dog and a little red wagon,
And a realio, trulio, little pet dragon.

Now the name of the little black kitten was Ink,
And the little grey mouse, she called him Blink,
And the little yellow dog was sharp as Mustard,
But the dragon was a coward, and she called him Custard.

Custard the dragon had big sharp teeth,
And spikes on top of him and scales underneath,
Mouth like a fireplace, chimney for a nose,
And realio, trulio daggers on his toes.

Belinda was as brave as a barrel full of bears,
And Ink and Blink chased lions down the stairs,
Mustard was as brave as a tiger in a rage,
But Custard cried for a nice safe cage.

Belinda tickled him, she tickled him unmerciful,
Ink, Blink and Mustard, they rudely called him Percival,
They all sat laughing in the little red wagon
At the realio, trulio, cowardly dragon.

Belinda giggled till she shook the house,
And Blink said Weeck! which is giggling for a mouse,
Ink and Mustard rudely asked his age,
When Custard cried for a nice safe cage.

Suddenly, suddenly they heard a nasty sound,
And Mustard growled, and they all looked around.
Meowch! cried Ink, and Ooh! cried Belinda,
For there was a pirate, climbing in the winda.

Pistol in his left hand, pistol in his right,
And he held in his teeth a cutlass bright,
His beard was black, one leg was wood;
It was clear that the pirate meant no good.

Belinda paled, and she cried Help! Help!
But Mustard fled with a terrified yelp,
Ink trickled down to the bottom of the household,
And little mouse Blink strategically mouseholed.

But up jumped Custard, snorting like an engine,
Clashed his tail like irons in a dungeon,
With a clatter and a clank and a jangling squirm,
He went at the pirate like a robin at a worm.

The pirate gaped at Belinda's dragon,
And gulped some grog from his pocket flagon,
He fired two bullets, but they didn't hit,
And Custard gobbled him, every bit.

Belinda embraced him, Mustard licked him,
No one mourned for his pirate victim.
Ink and Blink in glee did gyrate
Around the dragon that ate the pirate.

But presently up spoke little dog Mustard,
I'd have been twice as brave if I hadn't been flustered.
And up spoke Ink and up spoke Blink,
We'd have been three times as brave, we think,
And Custard said, I quite agree
That everybody is braver than me.

Belinda still lives in her little white house,
With her little black kitten and her little grey mouse,
And her little yellow dog and her little red wagon,
And her realio, trulio little pet dragon.

Belinda is as brave as a barrel full of bears,
And Ink and Blink chase lions down the stairs,
Mustard is as brave as a tiger in a rage,
But Custard keeps crying for a nice safe cage.

Ogden Nash

The Visitor

it came today to visit
and moved into the house
it was smaller than an elephant
but larger than a mouse

first it slapped my sister
then it kicked my dad
then it pushed my mother
oh! that really made me mad

it went and tickled rover
and terrified the cat
it sliced apart my necktie
and rudely crushed my hat

it smeared my head with honey
and filled the tub with rocks
and when i yelled in anger
it stole my shoes and socks

that's just the way it happened
it happened all today
before it bowed politely
and softly went away

Jack Prelutsky

123

The Terrible Path

While playing at the woodland's edge
I saw a child one day,
She was standing near a foaming brook
And a sign half-rotted away.

There was something strange about her clothes;
They were from another age,
I might have seen them in a book
Upon a mildewed page.

She looked pale and frightened,
Her voice was thick with dread.
She spoke through lips rimmed with green
And this is what she said:

'I saw a signpost with no name,
I was surprised and had to stare,
It pointed to a broken gate
And a path that led nowhere.

'The path had run to seed and I
Walked as in a dream.
It entered a silent oak wood,
And crossed a silent stream.

'And in a tree a silent bird
Mouthed a silent song.
I wanted to turn back again
But something had gone wrong.

'The path would not let me go;
It had claimed me for its own,
It led me through a dark wood
Where all was overgrown.

'I followed it until the leaves
Had fallen from the trees,
I followed it until the frost
Drugged the autumn's bees.

'I followed it until the spring
Dissolved the winter snow,
And whichever way it turned
I was obliged to go.

'The years passed like shooting stars,
They melted and were gone,
But the path itself seemed endless,
It twisted and went on.

'I followed it and thought aloud,
"I'll be found, wait and see."
Yet in my heart I knew by then
The world had forgotten me.'

Frightened I turned homeward,
But stopped and had to stare.
I too saw that signpost with no name,
And the path that led nowhere.

Brian Patten

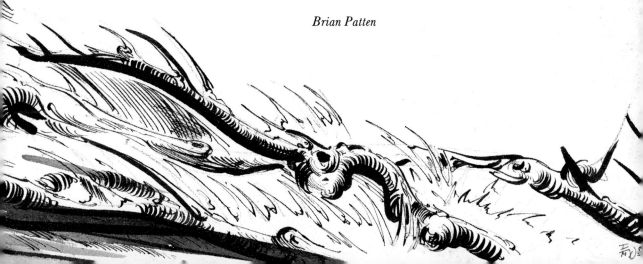

The Fairies

Up the airy mountain,
    Down the rushy glen,
We daren't go a-hunting
    For fear of little men;
Wee folk, good folk,
    Trooping all together;
Green jacket, red cap,
    And white owl's feather!

Down along the rocky shore
    Some make their home,
They live on crispy pancakes
    Of yellow tide-foam;
Some in the reeds
    Of the black mountain-lake,
With frogs for their watchdogs,
    All night awake.

High on the hill-top
    The old King sits;
He is now so old and grey
    He's nigh lost his wits.
With a bridge of white mist
    Columbkill he crosses,
On his stately journeys
    From Slieveleague to Rosses;
Or going up with music
    On cold starry nights,
To sup with the Queen
    Of the gay Northern Lights.

They stole little Bridget
   For seven years long;
When she came down again
   Her friends were all gone.
They took her lightly back,
   Between the night and morrow,
They thought that she was fast asleep,
   But she was dead with sorrow.
They have kept her ever since
   Deep within the lake,
On a bed of flag-leaves,
   Watching till she wake.

By the craggy hillside,
   Through the mosses bare,
They have planted thorn trees
   For pleasure, here and there.
Is any man so daring
   As dig them up in spite,
He shall find their sharpest thorns
   In his bed at night.

Up the airy mountain,
   Down the rushy glen,
We daren't go a-hunting
   For fear of little men;
Wee folk, good folk,
   Trooping all together;
Green jacket, red cap,
   And white owl's feather!

William Allingham

Sensitive, Seldom and Sad

Sensitive, Seldom and Sad are we,
As we wend our way to the sneezing sea,
With our hampers full of thistles and fronds
To plant round the edge of the dab-fish ponds;
Oh, so Sensitive, Seldom and Sad –
Oh, *so* Seldom and Sad.

In the shambling shades of the shelving shore,
We will sing us a song of the Long Before,
And light a red fire and warm our paws
For it's chilly, it is, on the Desolate shores,
For those who are Sensitive, Seldom and Sad,
For those who are Seldom and Sad.

Sensitive, Seldom and Sad we are,
As we wander along through Lands Afar,
To the sneezing sea, where the sea-weeds be,
And the dab-fish ponds that are waiting for we
Who are, Oh, so Sensitive, Seldom and Sad,
Oh, *so* Seldom and Sad.

Mervyn Peake

Nailsworth Hill

The Moon, that peeped as she came up,
   Is clear on top, with all her light;
She rests her chin on Nailsworth Hill,
   And, where she looks, the World is white.

White with her light – or is it Frost,
   Or is it Snow her eyes have seen;
Or is it Cherry blossom there,
   Where no such trees have ever been?

W. H. Davies

The Song of Wandering Aengus

I went out to the hazel wood,
Because a fire was in my head,
And cut and peeled a hazel wand,
And hooked a berry to a thread;
And when white moths were on the wing,
And moth-like stars were flickering out,
I dropped the berry in a stream
And caught a little silver trout.

When I had laid it on the floor
I went to blow the fire aflame,
But something rustled on the floor,
And some one called me by my name:
It had become a glimmering girl
With apple blossom in her hair
Who called me by my name and ran
And faded through the brightening air.

Though I am old with wandering
Through hollow lands and hilly lands,
I will find out where she has gone,
And kiss her lips and take her hands;
And walk among long dappled grass,
And pluck till time and times are done
The silver apples of the moon,
The golden apples of the sun.

W. B. Yeats

Who's That?

Who's that
stopping at
my door in the
dark, deep
in the dead of the moonless night?

Who's
that in the quiet
blackness,
darker than dark?

Who
turns the han-
dle of my door, who
turns the old brass hand-
le of
my door with never a sound, the handle
that always
creaks and rattles and
squeaks but
now
turns
without a sound, slowly
slowly
      slowly
          round?

Who's that moving through the floor
as if it were a lake, an open door? Who
is it who passes through
what can never be passed through,
who passes through
the rocking-chair
without rocking it,

who passes through
the table without knocking it, who
walks out of the cupboard without unlocking it?
Who's that? Who plays with my toys
with no noise, no
noise?

Who's that? Who is it
silent and silver
as things in mirrors, who's
as slow as feathers,
shy as the shivers,
light as a fly?

Who's that who's that
as close as
close as a hug, a kiss –

Who's THIS?

James Kirkup

The Old Wife and the Ghost

There was an old wife and she lived all alone
In a cottage not far from Hitchin:
And one bright night, by the full moon light,
Comes a ghost right into her kitchen.

About that kitchen neat and clean
The ghost goes pottering round.
But the poor old wife is deaf as a boot
And so hears never a sound.

The ghost blows up the kitchen fire,
As bold as bold can be;
He helps himself from the larder shelf,
But never a sound hears she.

He blows on his hands to make them warm,
And whistles aloud 'Whee-hee!'
But still as a sack the old soul lies
And never a sound hears she.

From corner to corner he runs about,
And into the cupboard he peeps;
He rattles the door and bumps on the floor,
But still the old wife sleeps.

Jangle and bang go the pots and pans,
As he throws them all around;
And the plates and mugs and dishes and jugs,
He flings them all to the ground.

Madly the ghost tears up and down
And screams like a storm at sea;
And at last the old wife stirs in her bed –
And it's 'Drat those mice', says she.

Then the first cock crows and morning shows
And the troublesome ghost's away.
But oh! what a pickle the poor wife sees
When she gets up next day.

'Them's tidy big mice,' the old wife thinks,
And off she goes to Hitchin,
And a tidy big cat she fetches back
To keep the mice from her kitchen.

James Reeves

133

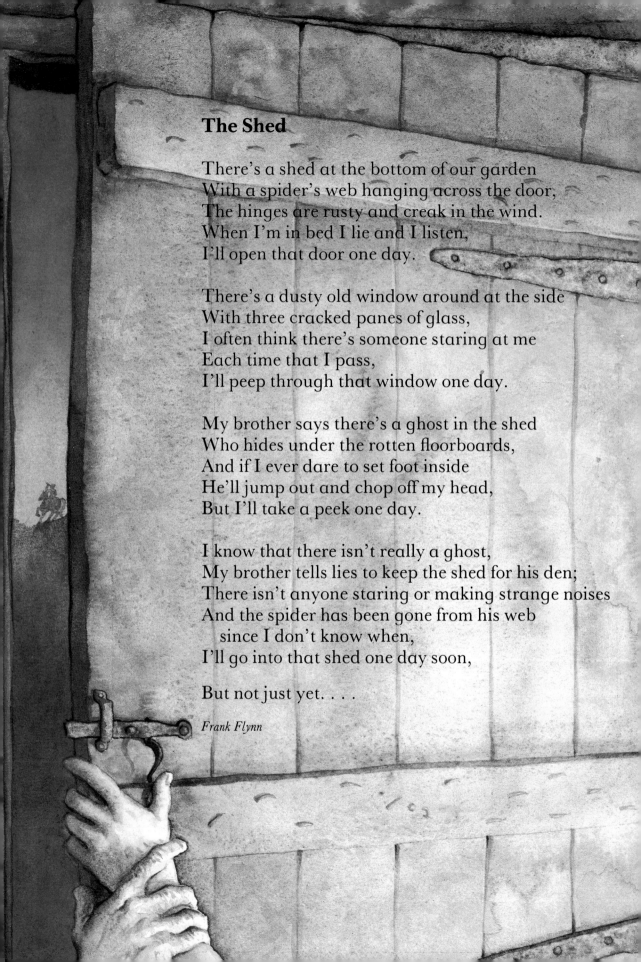

The Shed

There's a shed at the bottom of our garden
With a spider's web hanging across the door,
The hinges are rusty and creak in the wind.
When I'm in bed I lie and I listen,
I'll open that door one day.

There's a dusty old window around at the side
With three cracked panes of glass,
I often think there's someone staring at me
Each time that I pass,
I'll peep through that window one day.

My brother says there's a ghost in the shed
Who hides under the rotten floorboards,
And if I ever dare to set foot inside
He'll jump out and chop off my head,
But I'll take a peek one day.

I know that there isn't really a ghost,
My brother tells lies to keep the shed for his den;
There isn't anyone staring or making strange noises
And the spider has been gone from his web
    since I don't know when,
I'll go into that shed one day soon,

But not just yet. . . .

Frank Flynn

The Horn

'Oh, hear you a horn, mother, behind the hill?
My body's blood runs bitter and chill.
The seven long years have passed, mother, passed,
And here comes my rider at last, at last.
I hear his horse now, and soon I must go.
How dark is the night, mother, cold the winds blow.
How fierce the hurricane over the deep sea!
For a seven years' promise he comes to take me.'

'Stay at home, daughter, stay here and hide.
I will say you have gone, I will tell him you died.
I am lonely without you, your father is old;
Warm is our hearth, daughter, but the world is cold.'

'Oh mother, oh mother, you must not talk so.
In faith I promised, and for faith I must go,
For if that old promise I should not keep,
For seven long years, mother, I would not sleep.

Seven years my blood would run bitter and chill
To hear that sad horn, mother, behind the hill.
My body once frozen by such a shame
Would never be warmed, mother, at your hearth's flame.
But round my true heart shall the arms of the storm
For ever be folded, protecting and warm.'

James Reeves

The Late Express

There's a train that runs through Hawthorn
3 a.m. or thereabout.
You can hear it hooting sadly,
but no passengers get out.

'That's much too early for a train,'
the station-master said,
'but it's driven by Will Watson
and Willie Watson's dead.'

Poor Willie was a driver
whose record was just fine,
excepting that poor Willie
never learnt to tell the time.

Fathers came home late for dinner,
schoolboys late for their exams,
millionaires had missed on millions,
people changing to the trams.

Oh such fussing and complaining,
even Railways have their pride—
so they sacked poor Willie Watson
and he pined away and died.

Now his ghost reports for duty,
and unrepentant of his crime,
drives a ghost train through here nightly
and it runs to Willie's time.

Barbara Giles

The Garden Year

January brings the snow,
Makes our feet and fingers glow.

February brings the rain,
Thaws the frozen lake again.

March brings breezes, loud and shrill,
To stir the dancing daffodil.

April brings the primrose sweet,
Scatters daisies at our feet.

May brings flocks of pretty lambs
Skipping by their fleecy dams.

June brings tulips, lilies, roses,
Fills the children's hands with posies.

Hot July brings cooling showers,
Apricots and gillyflowers.

August brings the sheaves of corn,
Then the harvest home is borne.

Warm September brings the fruit;
Sportsmen then begin to shoot.

Fresh October brings the pheasant;
Then to gather nuts is pleasant.

Dull November brings the blast;
Then the leaves are whirling fast.

Chill December brings the sleet,
Blazing fire, and Christmas treat.

Sara Coleridge

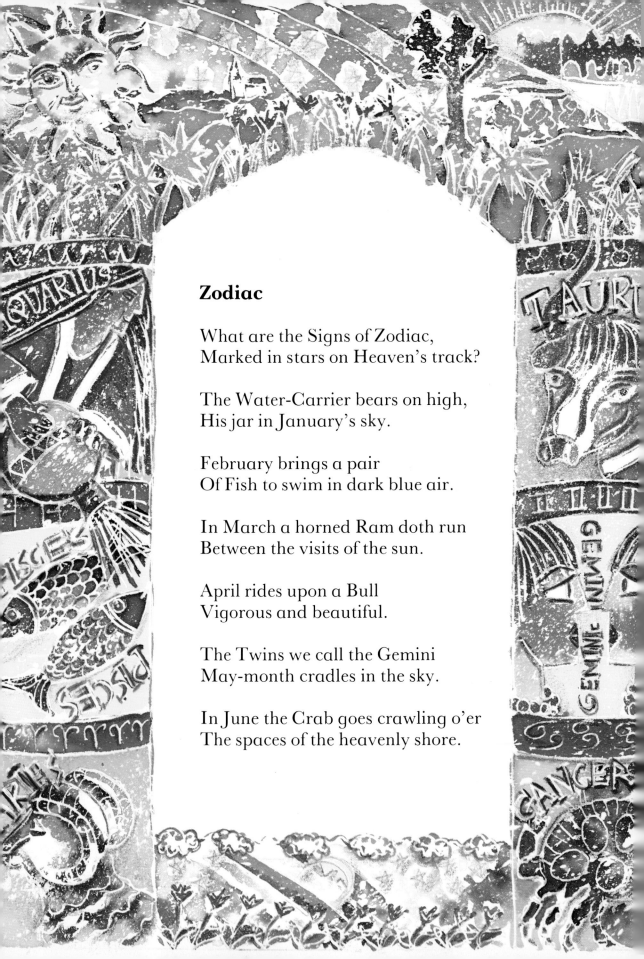

Zodiac

What are the Signs of Zodiac,
Marked in stars on Heaven's track?

The Water-Carrier bears on high,
His jar in January's sky.

February brings a pair
Of Fish to swim in dark blue air.

In March a horned Ram doth run
Between the visits of the sun.

April rides upon a Bull
Vigorous and beautiful.

The Twins we call the Gemini
May-month cradles in the sky.

In June the Crab goes crawling o'er
The spaces of the heavenly shore.

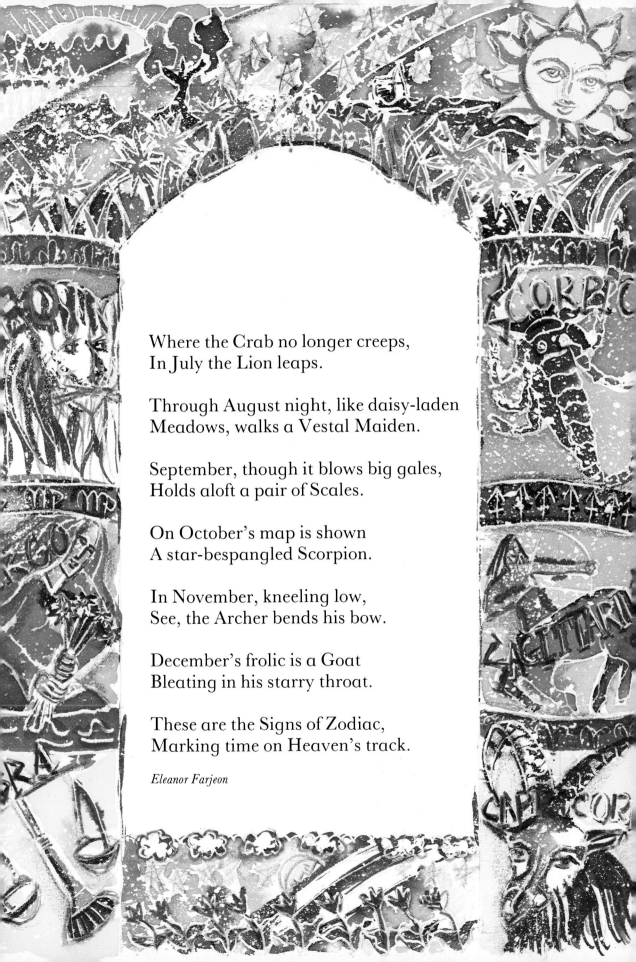

Where the Crab no longer creeps,
In July the Lion leaps.

Through August night, like daisy-laden
Meadows, walks a Vestal Maiden.

September, though it blows big gales,
Holds aloft a pair of Scales.

On October's map is shown
A star-bespangled Scorpion.

In November, kneeling low,
See, the Archer bends his bow.

December's frolic is a Goat
Bleating in his starry throat.

These are the Signs of Zodiac,
Marking time on Heaven's track.

Eleanor Farjeon

The Weather

What's the weather on about?
Why is the rain so down on us?
Why does the sun glare at us so?

Why does the hail dance so prettily?
Why is the snow such an overall?
Why is the wind such a tearaway?

Why is the mud so fond of our feet?
Why is the ice so keen to upset us?
Who does the weather think it is?

Gavin Ewart

Windy Nights

Whenever the moon and stars are set,
   Whenever the wind is high,
All night long in the dark and wet,
   A man goes riding by.
Late in the night when the fires are out,
Why does he gallop and gallop about?

Whenever the trees are crying aloud,
   And ships are tossed at sea,
By, on the highway, low and loud,
   By at the gallop goes he:
By at the gallop he goes, and then
By he comes back at the gallop again.

Robert Louis Stevenson

Clothes on the Washing Line

On windy days
Mum puts the washing on the line;
I think it's fun to watch
as she hangs Dad's shirts
upside down
and they wave their arms about
in a crazy sort of protest.
Mum's dresses always look
as though they're dancing,
but when I see my clothes
hanging on the line:
my favourite jeans
with patches on the knees,
my Liverpool football jersey
with a number seven on the back,
and a pair of grey football socks
that are supposed to be white,
it's like seeing bits of me
hanging there on the washing line.
I'm not really sure I like seeing
my clothes flapping in the wind,
I can't help feeling that I'm not altogether myself
and that I'm watching parts of me
waving me to join them.

Frank Flynn

141

When Skies are Low and Days are Dark

When skies are low
and days are dark,
and frost bites
like a hungry shark,
when mufflers muffle
ears and nose,
and puffy sparrows
huddle close –
how nice to know
that February
is something purely
temporary.

N. M. Bodecker

Storm

They're at it again
the wind and the rain
It all started
when the wind
took the window
by the collar
and shook it
with all its might
Then the rain
butted in
What a din
they'll be at it all night
Serves them right
if they go home in the morning
and the sky won't let them in

Roger McGough

Giant Thunder

Giant Thunder, striding home,
Wonders if his supper's done.

'Hag wife, hag wife, bring me my bones!'
'They are not done,' the old hag moans.

'Not done? not done?' the giant roars
And heaves his old wife out of doors.

Cries he, 'I'll have them, cooked or not!'
But overturns the cooking pot.

He flings the burning coals about;
See how the lightning flashes out!

Upon the gale the old hag rides,
The cloudy moon for terror hides.

All the world with thunder quakes;
Forest shudders, mountain shakes;
From the cloud the rainstorm breaks;
Village ponds are turned to lakes;
Every living creature wakes.

Hungry Giant, lie you still!
Stamp no more from hill to hill –
Tomorrow you shall have your fill.

James Reeves

143

Winter Morning

On cold winter mornings
When my breath makes me think
I'm a kettle,
Dad and me wrap up warm
In scarves and balaclavas,
Then we fill a paper bag
With bread and go and feed the ducks
In our local park.
The lake is usually quite frozen
So the ducks can't swim,
They skim across the ice instead,
Chasing the bits of bread
That we throw,
But when they try to peck the crumbs
The pieces slip and slide away.
Poor ducks!
They sometimes chase that bread
For ages and ages,
It makes me hungry just watching them,
So when Dad isn't looking
I pop some bread in my mouth and have a quick chew.
The ducks don't seem to mind,
At least they've never said anything
To me if they do.

Frank Flynn

144

The Snowman

Once there was a snowman
   Stood outside the door
Thought he'd like to come inside
   And run around the floor;
Thought he'd like to warm himself
   By the firelight red;
Thought he'd like to climb up
   On that big white bed.
So he called the North Wind, 'Help me now I pray.
   I'm completely frozen, standing here all day.'
So the North Wind came along and blew him in the door,
   And now there's nothing left of him
But a puddle on the floor!

Anon.

Winter Morning

Winter is the king of showmen,
Turning tree stumps into snow men
And houses into birthday cakes
And spreading sugar over lakes.
Smooth and clean and frosty white,
The world looks good enough to bite.
That's the season to be young,
Catching snowflakes on your tongue.

Snow is snowy when it's snowing,
I'm sorry it's slushy when it's going.

Ogden Nash

The Snowman

Mother, while you were at the shops
and I was snoozing in my chair
I heard a tap at the window
saw a snowman standing there

He looked so cold and miserable
I almost could have cried
so I put the kettle on
and invited him inside

I made him a cup of cocoa
to warm the cockles of his nose
then he snuggled in front of the fire
for a cosy little doze

He lay there warm and smiling
softly counting sheep
I eavesdropped for a little while
then I too fell asleep

Seems he awoke and tiptoed out
exactly when I'm not too sure
it's a wonder you didn't see him
as you came in through the door

(oh, and by the way,
the kitten's made a puddle on the floor)

Roger McGough

Ice

The North Wind sighed:
And in a trice
What was water
Now is ice.

What sweet rippling
Water was
Now bewitched is
Into glass:

White and brittle
Where is seen
The prisoned milfoil's
Tender green;

Clear and ringing
With sun aglow,
Where the boys sliding
And skating go.

Now furred's each stick
And stalk and blade
With crystals out of
Dewdrops made.

Worms and ants,
Flies, snails and bees
Keep close house guard,
Lest they freeze;

O, with how sad
And solemn an eye
Each fish stares up
Into the sky

In dread lest his
Wide watery home
At night shall solid
Ice become.

Walter de la Mare

White Fields

In winter-time we go
Walking in the fields of snow;

Where there is no grass at all;
Where the top of every wall,

Every fence and every tree,
Is as white as white can be.

Pointing out the way we came –
Every one of them the same –

All across the fields there be
Prints in silver filigree:

And our mothers always know
By the footprints in the snow,

Where it is the children go.

James Stephens

Escape at Bedtime

The lights from the parlour and kitchen shone out
    Through the blinds and the windows and bars;
And high overhead and all moving about,
    There were thousands of millions of stars.
There ne'er were such thousands of leaves on a tree,
    Nor of people in church or the Park,
As the crowds of the stars that looked down upon me,
    And that glittered and winked in the dark.

The Dog, and the Plough, and the Hunter, and all,
    And the star of the sailor, and Mars,
These shone in the sky, and the pail by the wall
    Would be half-full of water and stars.
They saw me at last, and they chased me with cries,
    And they soon had me packed into bed;
But the glory kept shining and bright in my eyes,
    And the stars going round in my head.

Robert Louis Stevenson

Is the Moon Tired?

Is the moon tired? She looks so pale
    Within her misty veil;
She scales the sky from east to west,
    And takes no rest.

Before the coming of the night
    The moon shows papery white;
Before the dawning of the day
    She fades away.

Christina Rossetti

Bedtime

Five minutes, five minutes more, please!
   Let me stay five minutes more!
Can't I just finish the castle
   I'm building here on the floor?
Can't I just finish the story
   I'm reading here in my book?
Can't I just finish this bead-chain —
   It *almost* is finished, look!
Can't I just finish this game, please?
   When a game's once begun
It's a pity never to find out
   Whether you've lost or won.
Can't I just stay five minutes?
   Well, can't I stay just four?
Three minutes, then? two minutes?
   Can't I stay *one* minute more?

Eleanor Farjeon

Bed in Summer

In winter I get up at night
And dress by yellow candle-light.
In summer, quite the other way,
I have to go to bed by day.

I have to go to bed and see
The birds still hopping on the tree,
Or hear the grown-up people's feet
Still going past me in the street.

And does it not seem hard to you,
When all the sky is clear and blue,
And I should like so much to play,
To have to go to bed by day?

Robert Louis Stevenson

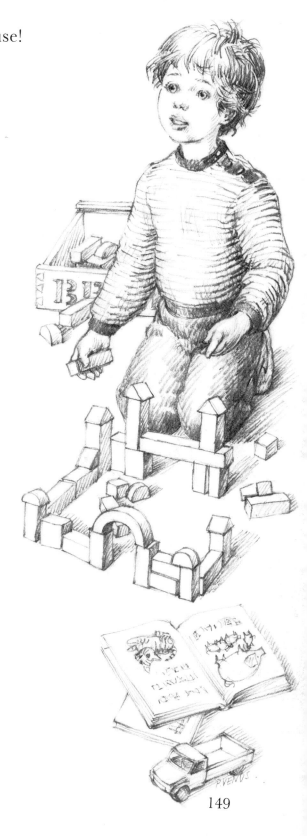

Bully Night

Bully night
I do not like
the company you keep
The burglars and the bogeymen
who slink
while others sleep

Bully night
I do not like
the noises that you make
The creaking and the shrieking
that keep me
fast awake.

Bully night
I do not like
the loneliness you bring
the loneliness you bring
The loneliness, the loneliness
the loneliness you bring,
the loneliness you bring
the loneliness, the

Roger McGough

I Often Meet a Monster

I often meet a monster
While deep in sleep at night;
And I confess to some distress.
It gives me quite a fright.
But then again I wonder.
I have this thought, you see.
Do little sleeping monsters scream
Who dream
Of meeting me?

Max Fatchen

Night Walk

What are you doing away up there
On your great long legs in the lonely air?
  Come down here, where the scents are sweet,
  Swirling around your great, wide feet.

How can you know of the urgent grass
And the whiff of the wind that will whisper and pass
  Or the lure of the dark of the garden hedge
  Or the trail of a cat on the road's black edge?

What are you doing away up there
On your great long legs in the lonely air?
  You miss so much at your great, great height
  When the ground is full of the smells of night.

Hurry then, quickly, and slacken my lead
For the mysteries speak and the messages speed
  With the talking stick and the stone's slow mirth
  That four feet find on the secret earth.

Max Fatchen

Skeleton House

Push, push the heavy door
CREE. . . CREE. . . CREEEEK!
Tip-toe the rotten floor
SQUEE. . . SQUEE. . . SQUEEEEK!
Step across the missing stair
EER. . . EEER. . . EEEERK!
Is that something over there?
SWISH. . . SWISH. . . SWISH. . .
Behind the curtain, what is that?
SCRITTER. . . SCRITTER. . . BUMP!
A red-eyed rat, a swooping bat
OOOW. . . OOOOOW. . . OOOOOW!
There's something sitting in that chair
SSH. . . SSSH. . . SSSSH!
His head is white with cobweb hair
OH!. . . NO!. . . SSH!
He starts to speak with clacking jaws
CLACK. . . CLACK. . . CLACK!
I grab his leg with all my force
PULL. . . PULL. . . . PULL. . . . PULL. .
Just like I'm pulling yours!

Laurence Smith

152

Whatif

Last night, while I lay thinking here,
Some Whatifs crawled inside my ear
And pranced and partied all night long
And sang their same old Whatif song:
Whatif I'm dumb in school?
Whatif they've closed the swimming pool?
Whatif I get beat up?
Whatif there's poison in my cup?
Whatif I start to cry?
Whatif I get sick and die?
Whatif I flunk that test?
Whatif green hair grows on my chest?
Whatif nobody likes me?
Whatif a bolt of lightning strikes me?
Whatif I don't grow taller?
Whatif my head starts getting smaller?
Whatif the fish won't bite?
Whatif the wind tears up my kite?
Whatif they start a war?
Whatif my parents get divorced?
Whatif the bus is late?
Whatif my teeth don't grow in straight?
Whatif I tear my pants?
Whatif I never learn to dance?
Everything seems swell, and then
The nighttime Whatifs strike again!

Shel Silverstein

Dragon Night

Little flame mouths,
Cool your tongues.
Dreamtime starts,
My furnace lungs.

Rest your wings now,
Little flappers,
Cave mouth calls
To dragon nappers.

Night is coming,
Bank your fire.
Time for dragons
To retire.

Hiss.
Hush.
Sleep.

Jane Yolen

Dragon

A dragon has come
To the side of my bed.
How did it get in?
It is, of course, red.

'O, jawsome, clawsome,
Scaily, flaily tailed beast
Do you want me
For a midnight feast?'

Turns off its fire.
Turns off its smoke.
'Dragon, did you hear
The words I spoke?'

Eyes me like my dog
When he sits up to beg
Then ambles away.
IT'S LAID AN EGG!

Olive Dove

The Troll to her Children

Billy Goat Gruff
Was yesterday's lunch,
So go to sleep fast
Or I'll give you a punch.

Jane Yolen

The Sleepy Giant

My age is three hundred and seventy-two.
And I think, with the deepest regret,
How I used to pick up and voraciously chew
The dear little boys whom I met.
I've eaten them raw, in their holiday suits;
I've eaten them curried with rice;
I've eaten them baked, in their jackets and boots.
And found them exceedingly nice.

But now that my jaws are too weak for such fare,
I think it exceedingly rude
To do such a thing, when I'm quite well aware
Little boys do not like to be chewed.

And so I contentedly live upon eels,
And try to do nothing amiss.
And I pass all the time I can spare from my meals
In innocent slumber – like this.

Charles Edward Carryl

155

Half Asleep

Half asleep
And half awake
I drift like a boat
On an empty lake.
And the sounds in the house
And the street that I hear
Though far away sound very clear.
That's my sister Betty
Playing by the stairs
Shouting like teacher
At her teddy bears.
I can hear Mum chatting
To the woman next door
And the tumble drier
Vibrates through the floor.
That's Alan Simpson
Playing guitar
While his Dad keeps trying
To start their car.
Dave the mechanic
Who's out on strike
Keeps revving and tuning
His Yamaha bike.
From the open window
Across the street
On the August air
Drifts a reggae beat.
At four o'clock
With a whoop and a shout
The kids from St John's
Come tumbling out.
I can hear their voices
Hear what they say
And I play in my head
All the games that they play.

Gareth Owen

Nurse's Song

Sleep, baby, sleep!
Your father herds his sheep:
Your mother shakes the little tree
From which fall pretty dreams on thee;
Sleep, baby, sleep!

Sleep, baby, sleep!
The heavens are white with sheep:
For they are lambs – those stars so bright:
And the moon's shepherd of the night;
Sleep, baby, sleep!

Sleep, baby, sleep!
And I'll give thee a sheep,
Which, with its golden bell, shall be
A little play-fellow for thee;
Sleep, baby, sleep!

Sleep, baby, sleep!
And bleat not like a sheep,
Or else the shepherd's angry dog
Will come and bite my naughty rogue;
Sleep, baby, sleep!

Sleep, baby, sleep!
Go out and herd the sheep,
Go out, you barking black dog, go,
And waken not my baby so;
Sleep, baby, sleep!

Anon.

Sweet Dreams

I wonder as into bed I creep
What it feels like to fall asleep.
I've told myself stories, I've counted sheep,
But I'm always asleep when I fall asleep.
Tonight my eyes I will open keep,
And I'll stay awake till I fall asleep,
Then I'll know what it feels like to fall asleep,
Asleep,
Asleeep,
Asleeeep . . . .

Ogden Nash

Good Night

Now good night.
Fold up your clothes
As you were taught,
Fold your two hands,
Fold up your thought;
Day is the plough-land,
Night is the stream,
Day is for doing
And night is for dream.
Now good night.

Eleanor Farjeon

158

At Night

I'm frightened at night
When they put out the light
And the new moon is white.

It isn't so much
That I'm scared stiff to touch
The shadows, and clutch

My blankets: it's – oh –
Things long, long ago
That frighten me so.

If I don't move at all,
The moon will not fall,
There'll be no need to call.

But, strangely, next day
The moon slips away,
The shadows just play.

Elizabeth Jennings

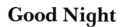

Good Night

Here's a body – there's a bed!
There's a pillow – here's a head!
There's a curtain – here's a light!
There's a puff – and so good night!

Thomas Hood

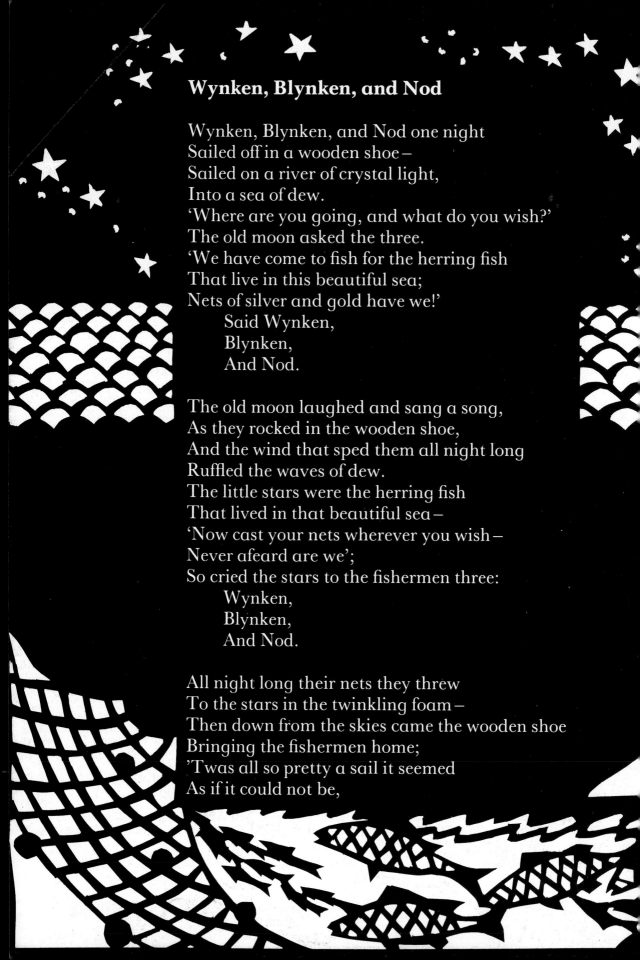

Wynken, Blynken, and Nod

Wynken, Blynken, and Nod one night
Sailed off in a wooden shoe—
Sailed on a river of crystal light,
Into a sea of dew.
'Where are you going, and what do you wish?'
The old moon asked the three.
'We have come to fish for the herring fish
That live in this beautiful sea;
Nets of silver and gold have we!'
    Said Wynken,
    Blynken,
    And Nod.

The old moon laughed and sang a song,
As they rocked in the wooden shoe,
And the wind that sped them all night long
Ruffled the waves of dew.
The little stars were the herring fish
That lived in that beautiful sea—
'Now cast your nets wherever you wish—
Never afeard are we';
So cried the stars to the fishermen three:
    Wynken,
    Blynken,
    And Nod.

All night long their nets they threw
To the stars in the twinkling foam—
Then down from the skies came the wooden shoe
Bringing the fishermen home;
'Twas all so pretty a sail it seemed
As if it could not be,

And some folks thought 'twas a dream they'd dreamed
Of sailing that beautiful sea –
But I shall name you the fishermen three:
    Wynken,
    Blynken,
    And Nod.

Wynken and Blynken are two little eyes,
And Nod is a little head,
And the wooden shoe that sailed the skies
Is a wee one's trundle-bed.
So shut your eyes while mother sings
Of wonderful sights that be,
And you shall see the beautiful things
As you rock in the misty sea,
Where the old shoe rocked the fishermen three:
    Wynken,
    Blynken,
    And Nod.

Eugene Field

The Man in the Moon

The Man in the Moon, as he sails the sky,
    Is a very remarkable skipper;
But he made a mistake when he tried to take
    A drink of milk from the Dipper.
    He dipped it into the Milky Way,
    And slowly and carefully filled it;
The Big Bear growled, and the Little Bear howled,
    And scared him so that he spilled it!

Anon.

The Star

Twinkle, twinkle, little star,
How I wonder what you are!
Up above the world so high,
Like a diamond in the sky.

When the blazing sun is gone,
When he nothing shines upon,
Then you show your little light,
Twinkle, twinkle, all the night.

Then the traveller in the dark,
Thanks you for your tiny spark,
He could not see which way to go,
If you did not twinkle so.

In the dark blue sky you keep,
And often through my curtains peep,
For you never shut your eye,
Till the sun is in the sky.

As your bright and tiny spark,
Lights the traveller in the dark —
Though I know not what you are,
Twinkle, twinkle, little star.

Jane Taylor

Star Wish

Star light, star bright,
First star I see tonight,
Wish I may
Wish I might,
Have the wish I wish tonight.

Anon.

Young Night Thought

All night long, and every night,
When my mamma puts out the light,
I see the people marching by,
As plain as day, before my eye.

Armies and emperors and kings,
All carrying different kinds of things,
And marching in so grand a way,
You never saw the like by day.

So fine a show was never seen
At the great circus on the green;
For every kind of beast and man
Is marching in that caravan.

At first they move a little slow,
But still the faster on they go,
And still beside them close I keep
Until we reach the town of Sleep.

Robert Louis Stevenson

163

From a Railway Carriage

Faster than fairies, faster than witches,
Bridges and houses, hedges and ditches,
And charging along like troops in a battle,
All through the meadows the horses and cattle:
All of the sights of the hill and the plain
Fly as thick as driving rain;
And ever again, in the wink of an eye,
Painted stations whistle by.

Here is a child who clambers and scrambles,
All by himself and gathering brambles;
Here is a tramp who stands and gazes;
And there is the green for stringing the daisies!
Here is a cart run away in the road,
Lumping along with man and load;
And here is a mill, and there is a river;
Each a glimpse and gone for ever!

Robert Louis Stevenson

Roadways

One road leads to London,
   One road runs to Wales,
My road leads me seawards
   To the white dipping sails.

One road leads to the river,
   As it goes singing slow;
My road leads to shipping,
   Where the bronzed sailors go.

Leads me, lures me, calls me
   To salt, green, tossing, sea;
A road without earth's road-dust
   Is the right road for me.

A wet road, heaving, shining,
   And wild with seagulls' cries,
A mad salt sea-wind blowing
   The salt spray in my eyes.

My road calls me, lures me
   West, east, south, and north;
Most roads lead men homewards,
   My road leads me forth.

To add more miles to the tally
   Of grey miles left behind,
In quest of that one beauty
   God put me here to find.

John Masefield

165

Travel

I should like to rise and go
Where the golden apples grow;
Where below another sky
Parrot islands anchored lie,
And, watched by cockatoos and goats,
Lonely Crusoes building boats;
Where in sunshine reaching out
Eastern cities, miles about,
Are with mosque and minaret
Among sandy gardens set,
And the rich goods from near and far
Hang for sale in the bazaar;
Where the Great Wall round China goes,
And on one side the desert blows,
And with bell and voice and drum,
Cities on the other hum;
Where are forests, hot as fire,
Wide as England, tall as a spire,
Where the knotty crocodile
Lies and blinks in the Nile,
And the red flamingo flies
Hunting fish before his eyes;
Where in jungles, near and far,
Man-devouring tigers are,
Lying close and giving ear
Lest the hunt be drawing near,
Or a comer-by be seen
Swinging in a palanquin;

Where among the desert sands
Some deserted city stands,
All its children, sweep and prince,
Grown to manhood ages since,
Not a foot in street or house,
Not a stir of child or mouse,
And when kindly falls the night,
In all the town no spark of light.
There I'll come when I'm a man
With a camel caravan;
Light a flower in the gloom
Of some dusty dining-room;
See the pictures on the walls,
Heroes, fights, and festivals;
And in a corner find the toys
Of the old Egyptian boys.

Robert Louis Stevenson

The Thin Prison

Hold the pen close to your ear.
Listen – can you hear them?
Words burning as a flame,
Words glittering like a tear,

Locked, all locked in the slim pen.
They are crying for freedom.
And you can release them,
Set them running from prison.

Himalayas, balloons, Captain Cook,
Kites, red brick, London Town,
Sequins, cricket bats, large brown
Boots, lions and lemonade – look,

I've just let them out!
Pick up your pen, and start,
Think of the things you know – then
Let the words dance from your pen.

Leslie Norris

Index of TITLES and first lines

First lines are shown in italic

Acknowledgements

The editors and publisher are grateful for permission to use the following copyright material:

John Agard: 'New Shoes', 'Lollipop Lady' and 'Ask Mummy, Ask Daddy', reprinted from *I Din Do Nuttin*, illustrated by Susanna Gretz, by permission of The Bodley Head. **N. M. Bodecker:** 'When Skies are Low and Days are Dark' reprinted from *Snowman Sniffles and Other Verse, copyright* © 1983 N. M. Bodecker (A Margaret K. McElderry Book), by permission of Faber & Faber Ltd., and Atheneum Publishers Inc. **Alan Brownjohn:** 'Seven Activities for a Young Child', reprinted by permission of the author. **Charles Causley:** 'Colonel Fazackerley', 'Old Mrs Thing-um-e-bob' and 'A Fox Came into my Garden' reprinted from *Figgie Hobbin* by permission of David Higham Assoc. Ltd. and Walker & Co. **Christine Chaundler:** 'The Tree in the Garden' reprinted from *The Golden Years* by permission of Robert Hale Limited. **Stanley Cook:** 'Boiling an Egg', reprinted by permission of the author. **W. H. Davies:** 'Nailsworth Hill', reprinted from *The Complete Poems of W. H. Davies*, © 1963 by Jonathan Cape Ltd., by permission of Jonathan Cape Ltd., on behalf of the Executors of the W. H. Davies Estate, and of Wesleyan University Press. **Walter de la Mare:** 'The Fly' and 'Ice' reprinted from *Complete Poems*, by permission of The Literary Trustees of Walter de la Mare and The Society of Authors as their representative. **Emily Dickinson:** 'I'm Nobody! Who Are You?' reprinted by permission of the publishers and the Trustees of Amherst College from *The Poems of Emily Dickinson*, edited by Thomas H. Johnson, Cambridge, Mass.: The Belknap Press of Harvard University Press. © 1951, 1955, 1979, 1983 by The President and Fellows of Harvard College. **Olive Dove:** 'Dragon', reprinted by permission of the author. **Gavin Ewart:** 'The Weather', reprinted by permission of the author. **Eleanor Farjeon:** 'Cats', 'Advice to a Child', 'Ned' and 'Zodiac' reprinted from *The Children's Bells*; 'Good Night' and 'Bedtime' reprinted from *Silver Sand and Snow*; 'Tailor' reprinted from *Then There Were Three*, by permission of David Higham Assoc. Ltd. **Max Fatchen:** 'It's a Bit Rich', 'I Often Meet a Monster' and 'Tailpiece' reprinted from *Wry Rhymes For Troublesome Times*; 'Night Walk' reprinted from *Songs For My Dog and Other People* by permission of John Johnson Ltd., and Penguin Books Ltd. **Aileen Fisher:** 'Upside Down' reprinted from *When It Comes to Bugs* and 'After a Bath' reprinted from *Up the Windy Hill*, both copyright of the author and reprinted with her permission. **F. Scott Fitzgerald:** extract from *The Crack-Up*, © 1945 by New Directions Publishing Corporation ('There was an orchestra-Bingo-Bango...'), reprinted by permission of The Bodley Head on behalf of the Estate of F. Scott Fitzgerald, and of New Directions Publishing Corporation. **Frank Flynn:** 'Winter Morning', 'The Shed' and 'Clothes on the Washing Line', © Frank Flynn 1984, reprinted from *The Candy-Floss Tree*: poems by Gerda Mayer, Frank Flynn, and Norman Nicholson (1984), by permission of Oxford University Press. **Esther Valck Georges:** 'Alley Cat' (Doubleday & Co Inc). **Barbara Giles:** 'Mrs Lorris, who Died of Being Clean' and 'The Late Express', © Barbara Giles 1983, reprinted from *Upright Downfall*: poems by Barbara Giles, Roy Fuller, and Adrian Rumble (1983) by permission of Oxford University Press. **Robert Heidbreder:** 'Copycat' from *Don't Eat Spiders* (Oxford University Press Canada 1985), used by permission of the publisher. **Russell Hoban:** 'Egg Thoughts' reprinted from *Egg Thoughts and Other Frances Songs*, text copyright © 1964, 1972 by Russell Hoban, by permission of Harper & Row, Publishers, Inc., and David Higham Associates Ltd.; 'The Friendly Cinnamon Bun', copyright © 1968 by Russell Hoban, reprinted from *The Pedalling Man* by permission of William Heinemann Ltd. and Harold Ober Associates Incorporated. **George Holloway:** 'Grown-ups' © 1987 Geoffrey Holloway reprinted by permission of the author. **Felice Holman:** 'Supermarket' reprinted from *At the Top of My Voice and Other Poems* (Scribners) **Patricia Hubbell:** 'Owl of the Greenwood', reprinted from *Catch Me A Wind*, copyright © 1968 Patricia Hubbell, by permission of Atheneum Publishers, Inc. **Elizabeth Jennings:** 'At Night' reprinted from *The Secret Brother* by permission of David Higham Associates Ltd. **James Kirkup:** 'Baby's Drinking Song' reprinted from *The Body Servant: Poems of Exile*; 'Who's That?' reprinted from *Ten and a Half* by permission of the author. **Robin Klein:** 'Amanda', reprinted from *Snakes and Ladders*, by permission J. M. Dent Pty. Limited. **Lois Lenski:** 'Sing a Song of People', copyright the Lois Lenski Covey Foundation. **Emily Lewis:** 'My Dog' reprinted from *Rainbow Lanterns* (Erskine Macdonald Ltd). **Hugh Lofting:** 'Picnic', copyright 1924 by Frederick A. Stokes & Co., renewed 1952 by Josephine Lofting, reprinted by permission of Christopher Lofting. **David McCord:** 'Mr Bidery's Spidery Garden' reprinted from *Mr Bidery's Spidery Garden* by permission of Harrap Limited, published in the United States in *For Me To Say* copyright © 1970 by David McCord, and reprinted from there by permission of Little, Brown and Company. **Roger McGough:** 'Storm' reprinted from *After the Merrymaking* by permission of Jonathan Cape Ltd; 'Harum Scarum', 'Potato Clock', 'The Snowman' and 'Bully Night' reprinted from *Sky In The Pie* (Kestrel Books), by permission of A. D. Peters & Co. Ltd. **John Masefield:** 'Roadways', reprinted from *Poems*, copyright 1912 by Macmillan Publishing Company, renewed 1940 by John Masefield by permission of The Society of Authors as the literary representative of the Estate of John Masefield and Macmillan Publishing Company. **Gerda Mayer:** 'Poor Mrs Prior', © 1984 by Gerda Mayer, first published on *Footnote 32* (Schools Poetry Association); 'Noah', © 1984 by Gerda Mayer, first published on *Footnote 33* (Schools Poetry Association), and 'Old Mrs Lazibones', first published in *The Knockabout Show* © 1978 by Gerda Mayer. All reprinted by permission of the author. **A. A. Milne:** 'Disobedience' and 'The King's Breakfast' reprinted from *When We Were Very Young*, Copyright 1924 by E. P. Dutton, renewed 1952 by A. A. Milne, by permission of The Publisher, E. P. Dutton, a division of NAL Penguin Inc, and of

Tailpiece

Tongues we use for talking.
Hands we clasp and link.
Feet are meant for walking.
Heads are where we think.
Toes are what we wiggle.
Knees are what we bend.
Then's there's what we sit on
And that's about the end.

Max Fatchen

Oxford University Press, Walton Street, Oxford OX2 6DP

Oxford New York
Athens Auckland Bangkok Bombay
Calcutta Cape Town Dar es Salaam Delhi
Florence Hong Kong Istanbul Karachi
Kuala Lumpur Madras Madrid Melbourne
Mexico City Nairobi Paris Singapore
Taipei Tokyo Toronto

and associated companies in
Berlin Ibadan

Oxford is a trade mark of Oxford University Press

This selection and arrangement © Michael Harrison
and Christopher Stuart-Clark 1988
First published 1988
Reprinted 1989, 1991, 1992, 1993
Reprinted in this format 1994, 1995
First published by Oxford in the United States in 1988
First published in paperback 1994
Reprinted 1995 (twice)

British Library Cataloguing in Publication Data

Oxford treasury of children' poems.
1. Children's poetry, English
I. Harrison, Michael, 1939   II. Stuart-Clark, Christopher
821'.914'0809282   PR1195.C47

Library of Congress Catalog Card Number 88-42849

ISBN 0 19 276055 6 (hardback)
ISBN 0 19 276134 X (paperback)

Set by Tradespools Ltd, Frome, Somerset

Printed in Hong Kong